The Music of the Heavens

The Music
of the Heavens

KEPLER'S HARMONIC

ASTRONOMY

Bruce Stephenson

PRINCETON UNIVERSITY PRESS

PRINCETON, NEW JERSEY

Copyright © 1994 by Princeton University Press
Published by Princeton University Press, 41 William Street,
Princeton, New Jersey 08540
In the United Kingdom: Princeton University Press, Chichester, West Sussex
All Rights Reserved

Library of Congress Cataloging-in-Publication Data

Stephenson, Bruce
The music of the heavens : Kepler's harmonic astronomy / Bruce Stephenson.
p.   cm.
Includes bibliographical references and index.
ISBN 0-691-03439-7
1. Planetary theory. 2. Kepler, Johannes, 1571–1630 Harmonices mundi.
I. Title.
QB361.S74   1994
521'.3—dc20   93-44916

This book has been composed in Adobe Utopia

Princeton University Press books are printed on acid-free paper
and meet the guidelines for permanence and durability of the
Committee on Production Guidelines for Book Longevity of the
Council on Library Resources

Printed in the United States of America

1 2 3 4 5 6 7 8 9 10

Speculationes de harmonia motuum coelestium ita succedunt ad votum exactissime et nobilissime, ut in summa voluptate dolere cogar, quod paucos admodum lectores fore praevideam.

—*Kepler to W. Schickard, 3 March 1618*

Contents

Preface and Ackowledgments

JOHANNES KEPLER'S REPUTATION is like the elephant in the poem "The Blind Men and the Elephant."[1] It has a very different feel depending upon which part of it one seizes on. Kepler is, of course, best known as an astronomer. His great astronomical works, the *Astronomia nova* and the *Epitome astronomiae Copernicanae* and the *Tabulae Rudolphinae*, give an unmistakable impression of solidity and importance, like an encounter with the "broad and sturdy side" of the elephant in the poem. His studies in other areas within the modern canon of the sciences, in mathematics and optics for example, are less monumental but equally secure. He explained how Galileo's telescope worked within a few months of learning that it existed. He confronted and solved many particular problems along the path that led Newton and Leibniz, half a century later, to the generalized methods of the calculus. He was one of the first to champion the use of logarithms as an aid to calculation.

Most of Kepler's fame is due to these irreproachably scientific aspects of his work. Other aspects of his reputation are less palatable today and hence less often discussed—except by writers wishing to portray the great scientist as a "mystic" or, more reasonably, a man as subject as the rest of us to the common ideas of his age. Kepler was, for instance, an astrologer. He accepted the traditional and nearly universal belief that the planets influenced both human affairs and the workings of nature, the weather in particular. He wrote not only the detailed astrological calendars that were a part of his official duties as a court mathematician but also a series of speculative and original works on the theoretical basis of astrological influence.

Most problematic of all are the works in which truth appears to emerge from a morass of delusion: above all, the *Mysterium cosmographicum* and the *Harmonice mundi*. The former was a youthful essay, written before Kepler attained any real competence as an astronomer. It can readily be understood for what it was, a paean to the Copernican system, whose merits had been imperfectly appreciated

---

[1] This short poem by J. G. Saxe is available in many anthologies of children's poetry, despite its concluding stanza in which the blind men are likened to theologians who "prate about an Elephant / Not one of them has seen!"

even by Copernicus himself. But the *Harmonice mundi* was a work of Kepler's maturity, one that he probably thought was more important than anything else he had written. It contained the first enunciation of his "harmonic law" relating the distances of the planets from the sun to their periods of revolution around it. Yet it radiated confidence, even in its most obscurely technical passages, that Harmony was to be found everywhere in the world, and that the heavens resounded (silently) to the same chords and scales used on Earth by musicians—whom Kepler therefore called the "apes of their Creator."

The strangeness of these ideas can be daunting. Yet Kepler expressed them with rational arguments, using words whose meaning is accessible. It is reasonable, therefore, for us to seek simply to understand what Kepler was trying to say to his readers. Until we have some grasp of what these words and arguments meant to him, our attempts to fit them into the larger context of early modern science will be clumsy and uninformed.

That is what the core of this book is about: to understand the fifth, astronomical book of the *Harmonice mundi* as Kepler wanted his readers to understand it. Chapter 9 will be particularly useful, I hope, to readers who are actually trying to follow Kepler's arguments—either in the original Latin or in English translation. The Latin text is available as volume 6 of the *Johannes Kepler Gesammelte Werke*[2] or in a facsimile edition such as the one listed in the bibliography. An imperfect but useable English translation of book 5 was included in the Great Books of the Western World series, volume 16.

The rest of the book should be of more general interest. The association of music with the heavens is very old and was once very respectable. It is now less prevalent—Kepler more or less exhausted the tradition, and the new astronomy he founded turned out, despite his hopes, to leave little room in the skies for harmony. I hope that my beginning chapters will provide enough context to make the *Harmonice mundi* comprehensible.

. . . . .

Noel Swerdlow first suggested that the very long article I was writing had grown to the point that I should think about turning it into a book. In the process of doing that, I have found his knowledge, both of astronomy and of music, to be invaluable. Since I first began working

---

[2] I will cite this edition with the abbreviation *G.W.*, giving volume, page, and (where appropriate) line numbers.

with him, he has been remarkably generous with his time. As the book neared completion, Tony Grafton tried to fill some of the larger gaps in my knowledge of the medieval and Renaissance periods and urged me to weigh the needs of my readers over my own interests. I have not always done what Noel and Tony suggested, but this would be a better book if I had.

My wife, Marija Norušis, has been (decreasingly) patient as my work on this book has stretched beyond reasonable limits. I am glad for both our sakes to be sending it to press.

I have held an appointment as visiting scholar in the Department of Astronomy and Astrophysics at the University of Chicago while preparing this book. I am grateful for this position and the access it provides to the University of Chicago library system. I would also like to acknowledge Columbia University's Rare Book and Manuscript Library, and Brad Westbrook in particular, for providing a microfilm of Offusius's rare treatise.

Figures 2.2, 5.1, 5.2, and 9.1 are reproduced by permission of the Department of Special Collections, University of Chicago Library. The Lassus font for the musical examples, copyright © 1991 by David Rakowski, is also used by permission.

The Music of the Heavens

Introduction

FOR AS LONG as people have contemplated the heavens, they have perceived music in the stately cycles of the motions overhead. Attempts to link the organized knowledge of astronomy and of music go back at least as far as Plato—as far as Pythagoras, if we heed the indistinct but nearly unanimous voice of tradition. Certainly Plato's account of the creation of the world was drenched in the language and symbolism of musical harmony. Music theorists of the Hellenistic world and encyclopedists of the Roman world expounded the subject, and through their writings it passed into the common knowledge of the learned in the middle ages. This tradition went well beyond a vague claim that the heavens seemed "harmonious" (in the sense of fitting together well) to make associations that were explicitly musical, and indeed rather technical, according to the best musical theories of the time. Often described poetically, the music of the heavenly spheres was also loosely integrated into the scientific view of the world as late as the seventeenth century. It was silenced, probably, by the increased precision with which astronomers measured the heavens, and probably also by the dissolution of the belief that Creation was specially arranged for the benefit and edification of humanity.

The last serious attempt, and perhaps the most earnest, to find musical harmony in the motions of the heavens was that of Johannes Kepler. Better known today for the laws of planetary motion he discovered, Kepler was a polymath who worked near the boundaries of contemporary knowledge in astronomy, mathematics, optics, physics, astrology, music theory, meteorology, and historical chronology. In most of these fields he searched more or less explicitly for the harmony that he was sure could be found in the world.

Harmony was for Kepler an ideal that was expressed in many ways, all based ultimately on the eternal relationships of geometry. The most important expressions of harmony were literally the biggest: the harmonies in the large-scale structure of the world and in the motions of its parts. His study of these celestial harmonies was closely related to his profound and original work as an astronomer. He believed that the heavens had been created, and the planets set into motion, as an embodiment of rational and harmonic principles. Kepler believed from an early age, before he acquired competence as an astronomer, that harmonic principles could be found in the heavens. His first book, the *Mysterium cosmographicum* of 1596, displayed to the world his discovery that the five "Platonic" solids were the key to the number and spacing of the planetary spheres. This insight into the world was geometric, and not specifically musical. While pursuing it as an assistant to Tycho Brahe, Kepler developed rather quickly into an astronomer of the first rank. Meanwhile, he did not neglect musical forms of celestial harmony. In 1599, he sent one of his correspondents a set of proportions that were harmonic and also, he thought, in remarkably close agreement with the best available sizes for the planetary orbits.

Over the next few years Kepler evidently drafted a substantial work on the harmony of the world, which he did not publish at that time. Finally in 1618, as the initial convulsions of the Thirty Years' War threw central Europe into turmoil around him, he discovered the relation between the sizes and periods of planetary orbits, a relation known today as his third, or harmonic, law of planetary motion. Working almost in a trance, as Max Caspar has suggested, he incorporated it into his harmonic theories, extending their scope greatly. The following year he published the work that he probably considered his greatest contribution to human knowledge.

HARMONICES MUNDI LIBRI V

The *Five Books of the Harmony of the World*,[1] which he published in 1619, treated harmony in its mathematical, musical, astrological, and astronomical aspects, culminating in an analysis of the harmonies in the motions of the planets. Treating all these diverse fields of knowledge under the single concept of "harmony" was not only permissible but necessary, in Kepler's opinion, for all these forms of harmony had the same mathematical basis. That basis was to be found in geometrical relations between physical quantities rather than in the purely numerical relations between integers on which Pythagorean harmonic theory was based. For a relation to be harmonic, Kepler

---

[1] *Harmonices mundi libri v* (Linz, 1619), vol. 6 in *G.W.* The word *Harmonices* is not a plural but a transliterated Greek genitive form, of which the nominative is *Harmonice.*

thought, its beauty had to be perceived by a soul, which might be human or otherwise. The planetary system, as revealed by Copernicus, expressed a harmony that was designed to be perceived at its center, by a soul in the Sun. Astronomy thus became the most precise and objective field for harmonic discovery. Many things on Earth were far removed from the hand of the Creator, but God alone had created the heavens: harmonies found there were inarguably divine.

In the *prooemium*, or preface, to book 5 of the *Harmonice mundi*, Kepler recalled his early conviction that the construction of the heavens was best understood from both geometric and harmonic principles. Before he had ever seen Ptolemy's *Harmonics*, he had convinced himself that the heavens had been built from a harmonic plan. Upon finally reading that book, in 1607, he had found, "beyond expectation and with astonishment, that almost all of [Ptolemy's] third book was devoted to this same contemplation of the Harmony of the heavens, fifteen hundred years ago."[2] Ptolemy's detailed account of these harmonies, based as it was on geocentric astronomy, and not much of that, was essentially useless to Kepler. This could hardly have been otherwise, he recognized. Ptolemy, ignorant of the heliocentric arrangement of the heavens, and equally ignorant of the theories of polyphonic harmony developed by Gioseffo Zarlino, Vincenzo Galilei, and others in the sixteenth century, could not possibly have understood the intricate cosmic harmonies. Kepler believed, in fact, that he was probably the first person in a position to grasp, in all its detail, the harmonic structure whose presence Ptolemy had faintly perceived in the heavens. His own work on astronomical harmony thus brought to fruition what Ptolemy had been unable to accomplish. The belief that he shared his inspiration with so great a scientist strengthened Kepler's determination to do correctly what the Alexandrine astronomer had been unable to complete.

The first two books of the *Harmonice mundi* dealt with the geometrical symmetries that gave rise, in Kepler's opinion, to all manifestations of harmony. J. V. Field has analyzed these difficult books admirably, highlighting the parts of this material that were mathematically original and pointing out the relevance of all the seemingly pure mathematics to Kepler's real concern, which was to use harmonic principles as an aid in understanding the natural world. The material, in fact, was not really pure mathematics but rather, in Field's words, "applica-

[2] *G. W.* 6.289.2–27.

ble mathematics." It was all intended as an aid in comprehending the harmonies found in nature.[3]

Book 3 of the *Harmonice mundi* dealt with specifically musical harmony, a subject about which Kepler knew a good deal more than is commonly realized. Kepler ranked the recent development of polyphonic theories of harmony alongside his own astronomical discoveries as a reason why it had not been possible in antiquity to give a true account of the harmonies in the sky. Although music theory is not a primary concern of ours, we will necessarily examine Kepler's concepts of *consonantia* and *concinna* intervals; of *durus* and *mollis*, concepts similar to but not quite the same as our major and minor; of *genus* and *modus* and *tonus* and all the attributes of "modern" sixteenth-century polyphony. Kepler found that all these things were included in the design of the heavens, and we cannot begin to appreciate that design as he did without learning to recognize them.

In his book 4, Kepler turned to astrology. He rejected a good part of traditional astrology but firmly believed in its fundamental principles. These he derived from the same harmonic considerations that underlay music and astronomy. Elsewhere, particularly in his professional capacity as court mathematician, Kepler wrote astrological prognostics in profusion. He collated ephemerides (daily tables of the calculated positions of heavenly bodies) with his own meteorological records and strove mightily to associate the vagaries of central European weather with the contemporaneous aspects of the heavenly bodies. He wrote several booklets, both in German and in Latin, to explain how planetary aspects affected the weather, the character of people born while the planets were in aspect, and the course of human affairs generally. Within this astrological oeuvre, book 4 of the *Harmonice mundi* occupies the position of a treatise on theoretical foundations. Kepler believed that astrological influence occurred when harmonic aspects among the heavenly bodies were perceived by souls, human or otherwise, on Earth. Meteorology was an important special case: the soul of the Earth itself was excited by aspects that it somehow was able to perceive, and in its excitement it exhaled vapors from its interior, which gave rise to winds, rain, and all kinds of storms.

The last of the five books making up the *Harmonice mundi* dealt with the astronomical harmonies of the world. It described in great detail the harmonic considerations involved in the large-scale struc-

---

[3] Field, *Kepler's Geometrical Cosmology* (Chicago, 1988), pp. 99–112.

ture of the world. These began, as a first approximation, with the nesting of "Platonic" regular polyhedra among the planetary spheres, to determine both the number and the sizes of those spheres. This theory had no musical content, but to Kepler it was nonetheless a harmonic theory. The regular polyhedra embodied the geometrical principles of abstract harmony, and he had discussed them already in book 2 as applications of those principles. Their appearance in the heavens illustrated the pervasiveness of harmonic proportions in the created world.

In the *Mysterium cosmographicum* Kepler had not really been able to resolve the discrepancies between the proportions implied by the polyhedral theory, on the one hand, and the relative distances of the planets according to astronomy, on the other. He had hoped that a proper reconstruction of heliocentric astronomy would clear up these problems. Since then, he had carried out the reconstruction himself; it had turned out to be much more thorough than he could have imagined beforehand. Yet it did not yield the distances he wanted, the distances implied by nesting the polyhedra among the planetary spheres. Accordingly, he summarized near the beginning of book 5 the state of the art in planetary astronomy—namely, his own theories—and then set out to show that the apparent motions of the planets, as perceived from the Sun, embodied all the proportions, all the variety, all the harmonic devices used to convey emotional nuance in contemporary polyphonic music. The mortal musicians who had invented all these things on Earth had been mere "apes of their Creator"; for the harmonic proportions involved existed eternally and had been embedded in the fabric of Creation.

Showing this much accomplished the goal Kepler had set for himself in his youth. The period-distance relation that he discovered while completing his *Harmonice mundi* enabled him to go much farther. In 1618 Kepler discovered this third or "harmonic" law, too late to weave it deeply into his tapestry. He stated it proudly enough in chapter 3 of book 5 but with only a few early and clumsy corollaries. He had not then decided (perhaps he never did decide) whether it was only an approximation, applying to orbits that were very nearly circular, or a fundamental result applying to all orbits whatever. In a few months of what must have been very intense labor, during the summer and fall of 1618, he managed to combine the new discovery with some of his other results, using it effectively at the end of book 5 to calculate the relative distances of the planets from the sun and showing thence that even the eccentricities of the individual planetary orbits were harmonically constrained to be what they were. The harmonic law bound together the different orbits and brought the book to a satisfying close. The whole planetary system was shown to be as beautifully constructed as it could possibly be. Kepler justified the few remaining dissonances not on aesthetic grounds—he was not so modern as that—but by showing that they were logically entailed by the other and more fundamental harmonies in the world.

Historians evaluating Kepler's harmonies have often overemphasized relatively trivial portions, such as the individual songs of the planets in chapter 6 of book 5, which make a nice illustration (see figure 9.9 in chapter 9). Important parts of the theory—the *latitudo tensionis*, or range of tuning in the universal harmonies, which permitted the harmonic configurations to occur far more frequently, and the highly desirable presence of extreme motions in those chords, which extended the duration of those harmonies—have been neglected. Above all, the roles played by Venus and the Earth, the "planets that change the type of harmony," have not been appreciated. The overall design made no sense if these roles were overlooked, for some of the actual proportions observed were excruciatingly unpleasant. Kepler explained such intervals by showing how they arose from the need to express harmonies of all possible types in the heavens.

Ironically, Kepler's revival of the ancient theory of celestial harmony came at just the time when a more exact knowledge of planetary distances and motions—which he himself had brought about—was making it nearly impossible to sustain the old theory. The *Harmonice mundi* was the final flowering of that theory, a fantastically detailed attempt to stretch the original idea to accommodate the New Astronomy that Kepler himself had created.

Kepler and Mysticism

In the three and a half centuries since Kepler published his *Harmonice mundi*, it has acquired a certain notoriety as the expression of an unscientific or mystical side of his personality. To be sure, Kepler's harmonic theories were fanciful, even for their time. In part this was simply due to his boldness as a thinker; he was not afraid to draw far-reaching conclusions where others, even if they agreed in principle, might lack the intellectual courage to follow him. A more important reason, I think, is that his theories were inspired by a theistic view of the world. This viewpoint could still be taken for granted in seventeenth-century Europe; it can no longer be taken for granted today. For most scientifically literate people in the twentieth century a blind mechanism—or something even stranger, such as probabilistic quantum mechanics—has replaced Kepler's Creator as the organizing principle of the world. Kepler's theories about the intimate details of Creation seem unscientific and irrational because they presuppose a created world in which humanity occupies a uniquely important place. This is, I think, the most important reason why his harmonic theories are condemned or praised (according to the taste of the critic) as "mystical."

Such descriptions are anachronistic and an abuse of language. Beliefs that are "mystical" are either held in opposition to reason or at least not subject to rational criticism. It may be possible to claim that any belief in a Creator, in the twentieth century, is inherently mystical; it is not possible for the seventeenth century. If Kepler was mystical because he believed that God had created the world and had created mankind in his image, then all of Europe espoused mysticism in the seventeenth century—and there is no point in using the word.

If the word *mystical* is to have any use in historical writing, it must be applied to ideas that are less rational, more transcendent, than the commonplace ideas of their time. The *Harmonice mundi* presented no ideas opposed to reason, nor any that claimed to transcend reason. The harmonies that Kepler discerned in the heavens were entirely rational. They had been created by divine reason; they were intended, Kepler assumed, for the enjoyment of a rational soul; and they were accessible to a sufficiently diligent exercise of human reason. Their foundations, after all, were in *geometry*, not in some vision vouchsafed to Kepler in a dream.[4] The strangeness of the *Harmonice mundi* today arises not from any lack of rationality in the discussion but from the diligence with which Kepler applied his reason to ideas that are not today deemed worthy of rational discussion. Its strangeness in Kepler's time arose from the dogged persistence with which he tried to reason about matters that his contemporaries thought to be beyond the reach of reason.

If these strange theories are not mystical, one is tempted to call

---

[4] Kepler did write a book about visions he supposedly had in a dream: the charming *Somnium*. This little book was partly learned jest and partly a lesson in how the heavens would appear from the Moon—a subject directly relevant to the dispute over heliocentrism. It contains nothing about celestial harmony.

them simply "wrong" and leave it at that. Their theoretical core is damaged beyond repair; it would be foolhardy to try to reconstruct science on the basis of a Creator's aesthetic choices. It is no simple matter, however, to proclaim a theory wrong. The geometrical relations that Kepler found are certainly not a part of living science. No modern scientist devotes time to them except out of curiosity, because they have no theoretical basis and because the empirical evidence for them is much less compelling than it appeared in 1619. The discovery of Uranus by Herschel in 1781, to say nothing of Neptune and Pluto in the nineteenth century, would have thrown the polyhedral theory into disarray if anybody had still been paying attention to it. A reconciliation of harmonic proportions to the precision with which we know planetary orbital parameters today, along the lines of what Kepler did in the thirty folio pages of chapter 9 of his book 5, would test the patience of a supercomputer.

Three and a half centuries ago these obstacles were smaller. Six planets were known; Kepler's theory not only exhibited harmonic proportions in their motions but explained why there were six. Galileo discovered four moons around Jupiter; Kepler promptly proposed that the nearly regular "rhombic solids" might explain their number and spacing just as the regular polyhedra explained those of the primary planets. He went on to show that the moons' orbits and periods agreed with his harmonic law.[5] His reconciliation of harmonic proportions, though dense reading, was probably about as precise as the precision of astronomical knowledge then required.

It is better to say that Kepler's harmonic theories were speculative. They identified what seemed to be striking patterns in the natural world and tried to explain those patterns. A useful analogy is the Titius-Bode law, which remains unexplained although it was first noticed in 1766 (aside from Uranus and the asteroids, which were unknown at the time but dutifully filled in gaps upon their discovery). The distances of the planets from the Sun are nearly in the proportions shown in table 1.1. The sequence finally breaks down for Neptune, although Pluto fits fairly well. Many attempts have been made to account for it, but it remains an unexplained regularity of nature.[6]

---

[5] See the fascinating (and well-illustrated) discussion of the rhombic solids in appendix 4 of Field. *Kepler's Geometrical Cosmology*. Kepler's discussion is in *G.W.* 4.309 and 7.318–19.

[6] For an introduction into the extensive literature on the Titius-Bode law, see M. Ovenden. "Bode's Law—Truth or Consequences," in A. and P. Beer, eds., *Vistas in Astronomy* (Oxford. 1975). vol. 18. pp. 473–96.

TABLE 1.1.
The Titius-Bode Law

1	2
Mercury	4
Venus	4+3
Earth	4+6
Mars	4+12
(asteroids)	(4+24)
Jupiter	4+48
Saturn	4+96
Uranus	4+192

Other unexplained regularities are not hard to find. Occasionally something unexpected befalls one of these orphans of science. The Balmer series was once nothing but an unexplained phenomenon observed in the spectrum of the hydrogen atom.[7] It is now revered as an important step, or sequence of steps, on the path to the Bohr atom.

It is unlikely that such an apotheosis awaits Kepler's harmonic theories after all these years. The harmonies of the spheres will continue to be heard as poetic metaphor but not as scientific truth. That is no reason for condescension. Among all those who have longed to hear those harmonies, Kepler was perhaps uniquely positioned—aware of the heliocentric system, having himself discovered the period-distance relation, but not yet burdened with an overly precise knowledge of the heavens—to conduct the last and most extravagant development of the Pythagorean theme.

STUDIES OF KEPLER'S ASTRONOMICAL HARMONIES

The standard edition of the *Harmonice mundi*, edited by Max Caspar, appeared as volume 6 of the *Gesammelte Werke*. Caspar's notes evince

---

[7] Absorption lines are observed at wavelengths of $h\frac{m^2}{m^2-2^2}$ where h is a constant and $m = 3, 4, 5, \ldots$

a fondness for this material. He gives more complete references for Kepler's literary allusions and provides algebraic proofs and paraphrases of many of Kepler's mathematical results.

Also of fundamental importance in the study of Kepler's astronomical harmonies is volume 5 of Frisch's *Opera omnia*, which contains, along with the first modern edition of the *Harmonice mundi*, Kepler's translation of Ptolemy's *Harmonics*, with related documents and correspondence. Frisch's notes are, as usual, uneven in coverage but always informed by a penetrating understanding of the text.

Among modern studies, J. V. Field's book *Kepler's Geometrical Cosmology* stands out for its analysis of all of Kepler's cosmological works. Field emphasizes the geometrical relations that Kepler believed to underlie all manifestations of harmony and devotes careful attention to the difficult problems involved in assessing the accuracy with which Kepler's harmonies fit the "facts" of astronomy—both his and ours. An account of the controversy between Kepler and Robert Fludd, a physician whose books are not readily available today, makes plain the differences between these two contemporary theorists of cosmic harmony. Any extended discussion of these subjects in my book has been rendered superfluous by Field's. The illustrations, reproduced from the original editions, provide a good bit of the flavor of Kepler's books.

Treatment of musical astronomy is uneven in other standard accounts of Kepler's work. Dreyer's classic study of the history of astronomy devotes half a dozen pages to what he regarded as the essence of Kepler's theories. They are accurate generally but not always specifically. Max Caspar's biography, *Kepler 1571–1630*, addresses all the important questions, based on Caspar's broad familiarity with Kepler's writings. The few pages devoted to the *Harmonice mundi* cannot explain any of the details but leave a warm and rather vague impression of Kepler as a man of great spirit, clear mind, and profound philosophy. A. Koyré in *The Astronomical Revolution* translates a long passage from the summary of astronomy toward the beginning of book 5 of the *Harmonice mundi* and reproduces several of the more important tables from the end. His discussion and notes are not fully thought out, as is often true in this book, and raise as many questions as they resolve.[8]

---

[8] Koyré, *The Astronomical Revolution* (Ithaca, 1973), sec. 2, part 3, chap. 2 (entitled "The *Harmonice mundi*"). This book perhaps exceeds all others in the difficulty one has in finding a note from a reference in the text.

On specifically musical aspects of Kepler's work I have found two studies to be particularly valuable. The Renaissance historian D. P. Walker has written a brief but thorough analysis of the musical content of Kepler's celestial harmonies. M. Dickreiter has written a book-length monograph on Kepler as musical theorist, based primarily on book 3 of the *Harmonice mundi*, which has (I hope) compensated somewhat for my relative ignorance of Renaissance musical theory.[9] I do not doubt that infelicities remain in my own discussion and can plead only that my chief interest is in Kepler's application of music to the heavens.

THE PLAN OF THIS BOOK

Chapter 2 of this book takes a broad look at ancient theories associating music with the planets. A number of such theories were quite specific about the intervals between the planets' notes, but we know little about the significance of such details. The most important ancient theory was that of Ptolemy, the astronomer, who alluded to planetary music in the Canobic Inscription and described it in the last three chapters of his *Harmonics*. Much from these chapters has been lost, unfortunately, but one can infer some of what they probably contained. The knowledge that Ptolemy had written on this subject inspired Kepler to do the job properly, with his more "modern" knowledge of both astronomy and music theory. The chapter concludes with a survey of *musica mundana*—world music, as this material was called—between Ptolemy and Kepler.

Chapter 3 gives an account of the data on planetary distances provided first by geocentric and then by heliocentric astronomy, to show how the target distances for harmonic theories had changed so radically in Kepler's age. In chapter 4, I turn to the work of a little-known sixteenth-century astrologer, Jofrancus Offusius. Like many others of his time, Offusius wanted to flush out the nonsense that had accumulated in astrological theory and practice. His prescriptions for reform were a greater respect for observation and a mathematical analysis of the circumstances under which the planets exerted their influence. For us his importance is in the mathematical patterns he postulated, not only among the distances to the planets but also among the ec-

---

[9] Walker, "Kepler's Celestial Music"; Dickreiter, *Der Musiktheoretiker Johannes Kepler* (Bern and Munich, 1973).

centricities of their orbits and the strengths with which they radiated the four basic qualities of heat, cold, moisture, and dryness. Although the patterns proposed by Offusius were superficially similar to those in which Kepler believed, their author subjected them to neither the rigorous analysis nor the withering empirical assessment that combine to make the *Harmonice mundi* the finest example of its genre.

In chapter 5 we take up the work of Kepler himself, looking first at the *Mysterium cosmographicum* and its polyhedral theory of planetary distances. Chapter 6 finds Kepler three years later, proposing his first theory about the musical harmonies of the heavens in letters to (among others) his patron Herwart von Hohenburg and his former teacher Michael Maestlin. As had been the case with the polyhedral theory, he is unable to make this first harmonic theory work properly but hopes that he will be able to confirm it by obtaining more accurate parameters for the planetary orbits.

Obtaining those parameters required many long years of work, as it turned out, for in the process Kepler had to create a new astronomy, whose physically based principles were unprecedented in the long history of that science. After completing the essential work, which he published in the accurately titled *Astronomia nova* of 1609, he was at last able to devote more time to his beloved harmonic speculations. He obtained a manuscript that included most of the Greek text of Ptolemy's *Harmonics* and prepared a translation of it, with extensive notes explaining, criticizing, and correcting the theories of his esteemed predecessor. Kepler never published his translation or notes, but they remained with his papers at his death. In chapter 7 we take a close look at Kepler's necessarily creative interpretation of Ptolemy's work on astronomical harmony.

Kepler finally published his own theories, covering all aspects of harmony in the world, in 1619 as *Harmonices mundi libri v*, or "Five Books of the Harmony of the World." The core of the present work, in chapters 8 and 9, is a detailed analysis of Kepler's fifth book, wherein he treats the harmonies of the astronomical world. These harmonies were probably the subject that interested him most, of all his many interests. They have rightly received a great deal of attention since he published them—most of it, unfortunately, either patronizing or adulatory. They were by no means as naive as one might expect from common accounts. Kepler himself was fully aware that his harmonic propositions were not precisely true but had of necessity been balanced against other potential truths by what Curtis Wilson has called "the complex artistry of the Creator."[10]

A good deal of Kepler's own artistry in the *Harmonice mundi* was to discover the competing (because incompatible) claims to harmony of various arrangements and explain the principles by which these claims had been resolved by the Creator. His chief principle was variety: variety of harmonic intervals, of course, but also variety in the genus of the intervals. In the lengthy chapter 9 of book 5, he derived most of the difficult intervals from the roles of Earth and Venus as the planets that acted out a marital dance around the Sun, switching the great silent chords of celestial harmony from major to minor and back. (I am oversimplifying here; all will be made clear, I hope.) The astonishing variety of harmony revealed by Kepler's analysis in book 5 must have given endless delight to its author—very much in the same way he thought it had given delight to the Author of the world.

My concluding chapter takes a look at a couple of seventeenth-century writers, an English astronomer and an Italian Jesuit teacher, who read the *Harmonice mundi* with understanding. I close by arguing what I hope will be clear to anyone who reads this book: that Kepler was not a mystic, as is often claimed, but rather a man of his age, devout and rational at the same time.

---

[10] C. Wilson, "Horrocks, Harmonies, and the Exactitude of Kepler's Third Law," p. 241. Wilson argued that by the mid-1620s Kepler had concluded that his laws did not always apply exactly, although they were fundamentally correct. I think this was his attitude from the beginning, and in chap. 9 I will point out how already in the *Harmonice mundi* he qualified the third law, as if unsure to what extent it was valid.

Earlier Theories of Astronomical Harmony

BEFORE TURNING our attention to Kepler, let us take a look at the tradition that climaxed in his *Harmonice mundi*. The tradition probably started with the practical observation that bodies rubbing together make noise. The heavenly spheres, which were enormous and which moved with different speeds, would by this reasoning surely make a tremendous racket. This account was sufficiently old that it was credited to Pythagoras by most writers. Aristotle describes an ancient version of the theory, attributing it only to "the Pythagoreans."[1]

According to Aristotle these Pythagoreans asserted that the circular rotations of the heavenly spheres produced sounds that were surely very loud. Moreover their speeds, "as measured by their distances," were in the same ratios as musical concords, so the sounds themselves were surely harmonies. The Pythagoreans accounted for our inability to hear such great sounds by asserting that we would hear them, except that they have been with us every moment since birth. Having never experienced the absence of the celestial harmonies, we do not notice their presence.

Aristotle himself does not think much of this theory. The noise is never heard, he states, for the simple reason that there is no noise. If the heavenly spheres were forcibly moved through surrounding regions different from themselves, we would surely hear (and be shattered by) the noise of their motion. Since we do not hear, and are not shattered, he concludes that the celestial spheres are not forced into motion. He does not comment on the assumption that the planets' distances are in harmonic proportions.

PLATO'S CELESTIAL MUSIC

Two passages in Plato's writings provided the most authoritative source from which medieval Europe learned of heavenly music. One

---

[1] *De caelo*, 290B–91A; pp. 479–80 in the Barnes edition.

was the myth of Er, in the tenth book of the *Republic*.[2] Er was a warrior slain in battle, who supposedly came to life on his funeral pyre and recounted what his spirit had seen in the otherworld. Along with the terrible spectacle of souls being judged, he described a vision of eight concentric whorls in the heavens. Each whorl carried a siren, "uttering one sound, one note, and from all the eight there was the concord of a single harmony." The musical content of this passage is limited to a vision of sirens singing as they are carried around by the heavens; there is no attempt to develop the theme. Much more important in later accounts was the story of creation in the *Timaeus*.

The *Timaeus*,[3] probably more than any other of Plato's dialogs, influenced later attempts to understand the natural world. Its mixture of poetry, metaphysics, and mathematics offered an apparently scientific—if not always comprehensible—basis for understanding the world. According to Timaeus, the narrator, God desired that all things should be good, so far as this was possible. He built a creature, the world, in the form of a sphere, which was the most perfect shape possible. To it he gave the best and most suitable kind of motion, which was rotation about its center, without giving it any motion from place to place. As ingredients for the world, there were three different kinds of stuff (one is really not certain how to categorize such abstractions): Being, the Same, and the Different. Each of the three existed both as an indivisible whole and as something that is divisible and shared among different bodies. From these two extreme manifestations of each of the three concepts, the Creator compounded three intermediates: Being, the Same, and the Different, each of which was partly indivisible and partly divided.

This was not yet the world we know. The Creator combined the three new intermediate elements (apparently using force to unite the Same with the Different). When he had formed a single amalgam, compounded from all the abstract opposites, he began to work Mathematics upon it. He took away 1 part; then he took away 2 parts and 3 parts; then 4 parts and 9 parts; then 8 parts and 27 parts. Thus were formed the series of double numbers (1, 2, 4, 8) and the series of triple numbers (1, 3, 9, 27). These numbers, the squares and the cubes, combined to form the sequence 1, 2, 3, 4, 9, 8, 27, a number sequence that, according to the Pythagorean point of view that Timaeus expounds

---

[2] *Republic*, 614b–21b; pp. 839–44 in the Hamilton and Cairns edition.

[3] I have used Jowett's translation, from the Hamilton and Cairns collection.

here, was of the utmost importance in determining the structure of the world.

To this point the story deals with numbers; but these numbers, as Timaeus explains, contain all that is essential in musical harmony. Within the intervals of each series, the Creator inserted new numbers that were the harmonic and arithmetic means of adjacent pairs of numbers. Between 1 and 2 the harmonic mean was $\frac{4}{3}$, and the arithmetic mean $\frac{3}{2}$. These two means in music represent the intervals of a fourth and a fifth, respectively.[4] The difference between these intervals was $\frac{9}{8}$, which represents a whole tone. Taking this new interval of $\frac{9}{8}$ as a unit, the Creator removed two such units out of the interval of a fourth, $\frac{4}{3}$, and found that the remaining small interval was in the ratio $\frac{256}{243}$. In this fashion he divided harmonically the entire mixture of combined opposites with which he had begun.

Next, the Creator sliced the harmonically divided compound in half lengthwise into two strips, laying one strip across the other in the form of an X. He then joined together the two ends of each strip to form a pair of circles that crossed twice, once at the crossing of the original X and once opposite it, where the four ends came together. One of these circles was outer and one inner; and the Creator set each of them into motion around its center. The outer circle moved to the right. Its motion was simple and undivided and was given dominion over that of the inner circle. One can easily identify this outer circle with the celestial equator, whose westward motion is "to the right" for any observer in the northern hemisphere.

The inner circle, on the other hand, was further divided into seven unequal circles. These seven circles were separated by six intervals "in ratios of two and three, three of each," and set into motion in the opposite direction. Three of the seven were made to move with equal swiftness, while the other four all moved at different speeds "but in due proportion." These backward-moving circles were the seven planets of antiquity, which moved eastward or to the left against the dominating daily rotation of the celestial equator. The Sun, Mercury, and Venus moved with equal swiftness, completing one rotation per year on average, whereas the Moon, Mars, Jupiter, and Saturn each

---

[4] A ratio represents a musical interval if two strings, of identical composition, thickness, and tension, but of lengths in the given ratio, sound the interval when plucked. Identical strings of lengths 4 and 3 sound the interval of a fourth; identical strings of lengths 3 and 2 sound the interval of a fifth.

had a speed of its own. The order of planets above the Earth was Moon, Sun, Venus, and Mercury, followed by the others.[5]

The odd phrase describing the intervals between the seven circles, "in ratios of two and three, three of each," is certainly a reference to the sequences 1, 2, 4, 8 and 1, 3, 9, 27. Quite possibly it implies that the sizes of the seven circles are proportional to the seven numbers in those sequences, or something similar. (We will see a different interpretation in Macrobius, presently.) If this is so, the Platonic distances—or whatever they are—describing the harmony of the planets are as shown in table 2.1.

Timaeus goes on to narrate poetically the manner of creation of all the things inhabiting the world. Eventually he gets around to describing an abstract but surprisingly elaborate theory of matter based on atomic units that are shaped like tiny polyhedra. The description is, typically, given in purely geometric terms, without reference to "real" matter: for Plato, mathematical forms *are* reality. The building blocks of matter are the regular polyhedra, which themselves are formed from regular polygons, which in turn are ultimately pieced together from two types of tiny triangles.[6]

One type of fundamental triangle is the 30°–60°–90° right triangle formed by splitting an equilateral triangle down the middle. Six of these, arranged symmetrically, form an equilateral triangle (figure 2.1). Four equilateral triangles can be assembled into a regular tetrahedron; eight of them can be assembled into a regular octahedron; and twenty of them can be assembled into a regular icosahedron. The other type of fundamental triangle is the 45°–45°–90° right isosceles triangle (also in figure 2.1). It is used to build a square symmetrically. Six squares can then be assembled into a cube, the fourth of the regular polyhedra. Plato remarks that a fifth such figure exists, "which God used in the delineation of the universe with figures of animals." He does not attempt to construct it from his fundamental triangles, since this is not possible.

From these five "Platonic solids," or regular polyhedra, the Creator made the elements out of which the world was constructed. Earth was made from the cubic form, fire from the tetrahedron, air from the octahedron, and water from the icosahedron. The dodecahedral

---

[5] *Timaeus*, 35A–38D; pp. 1165–68 in the Hamilton and Cairns edition.

[6] This and the following are from *Timaeus*, 54C–56B; pp. 1180–82 in Hamilton and Cairns.

TABLE 2.1.
Platonic Harmonies

1	2	3
Planet	Distance (?) above Earth	Interval from preceding planet
Moon	1	
Sun	2	Octave
Venus	3	Fifth
Mercury	4	Fourth
Mars	8	Octave
Jupiter	9	Tone
Saturn	27	Octave+fifth

shape was used to make the fifth element, of which the heavens (including the zodiac, with its figures of animals) were composed.

Plato's associations of the regular polyhedra with the elements became firmly established in Western iconography. Figure 2.2 shows the polyhedral figures as Kepler printed them in the *Harmonice mundi*,[7] each adorned with an image suitable for its element. The cube has earthy pictures on its faces: a tree with what appears to be a harrow beside it, a pair of crossed farm implements, and a carrot or some other root vegetable. The tetrahedron contains a pair of sticks blazing away in a little fire. The octahedron depicts the element of air with three birds flying among the clouds, although one bird actually appears to be sitting in midair; and the watery icosahedron is decorated with a lobster and a couple of leaping fishes. The dodecahedron, most impressively emblematic of the lot—it is reproduced on the cover of each volume of Kepler's *Gesammelte Werke*—displays a fiery sun, with an amiably stupid-looking face, surrounded by stars and a crescent moon or planet.

The account of the creation of the heavens given by *Timaeus* leaves

[7] G.W. 6.79.

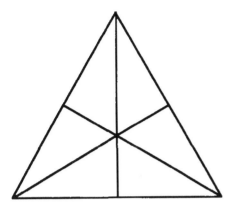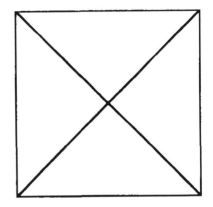

FIGURE 2.1. Plato's construction of the equilateral triangle and square

much unresolved. Its description of the foundations of our world, however, which resonated through Western thought into the modern age, was permeated with musical proportions. The original compound of primordial qualities, from which all the world was made, had been divided into the harmonic intervals even before it was used to construct the heavens. (This "division" was presumably done as in the fifth book of Euclid, with compass rather than blade.) The seven circles carrying the seven planets were separated by intervals in the harmonically constructed proportions 1, 2, 3, 4, 8, 9, 27. At any rate, the intervals between the planets' circles somehow formed harmonic proportions. Finally, the motions of the planets' circles, where unequal, were "in due proportion." The stuff of the world had been measured harmonically, cut up harmonically, and set into motion harmonically.

Plato's *Timaeus* introduced celestial harmony into the very heart of Western natural philosophy, where it has not lacked students and commentators. I shall not attempt to make an original contribution on the topics discussed by so many eminent philosophers.[8] Instead I will follow the thread of specifically musical associations with the planets, a thread that ultimately leads to Kepler and his *Harmonice mundi*. Plato's harmonies, quite clearly, were dictated by rational and aesthetic considerations and expressed poetically with little regard to how well they agreed with possible measurements or observations. Even

---

[8] For example, see the commentaries of Taylor and of Cornford. The first chapter of J. V. Field's *Kepler's Geometrical Cosmology* (Chicago, 1988) analyzes Plato's polyhedral theory of matter with an eye toward Kepler's later use of the regular polyhedra.

FIGURE 2.2. The regular polyhedra and their elements

Kepler's most imaginative theories grew from his knowledge of the world, and he tested them constantly against that knowledge. He admired the Platonic philosophy, but his work as a scientist had nothing to do with Plato.

Musical Astronomy in the Hellenistic World

Ancient Greek music theorists may be divided loosely into two schools, the "Pythagoreans" and the school of Aristoxenus. Pythagorean harmony, for which the *Timaeus* is an important source, was based on the arithmetic sequence 1, 2, 3, 4 and considered only the ratios among these numbers to be consonances:

2:1 or 4:2	Octave	*Dia pason*
3:2	Fifth	*Dia pente*
4:3	Fourth	*Dia tessaron*
3:1	Fifth + octave	*Dia pente kai dia pason*
4:1	Double octave	*Dis dia pason*

The tone was defined as the difference between a fifth and a fourth and hence corresponded to a ratio of 9:8. Theorists of the Pythagorean school derived musical concepts directly from mathematics and in particular from arithmetic. The rival school of Aristoxenus (fourth century B.C.) rejected any arithmetic basis for music. Aristoxenus and his followers preferred to treat notes as themselves fundamental rather than make them dependent on the more general concepts of mathematics.

Associations of individual planets with the notes of the musical scale were made, naturally enough, by writers of the Pythagorean school and were usually attributed to Pythagoras himself. Von Jan, who published the basic modern study of Greek theories on the harmony of the spheres,[9] argued that Pythagorean knowledge of astronomy was not sufficient for such "theories" to consist of more than facile analogies. (For example, five regular polyhedra matched up nicely with the five elements, if one included a "fifth essence" making up the immutable heavens.) Little is known with any certainty of authentic Pythagorean doctrines.

Early Planetary Scales

The earliest surviving theories of planetary harmony assign planets not to notes in a chord but rather to notes in a series or scale, whose harmonic character was realized sequentially rather than simultaneously. The intervals between notes are simply associated one-to-one with the distances between the heavenly spheres. This may or may not have been significant to the Greek attitude toward the harmony of the spheres. There is no question, as we shall see, that Kepler thought it a crippling limitation to the ancient theories. To him, monophonic music was as inadequate a representation of the celestial harmonies as geocentric astronomy was of the celestial motions.

Pliny, that amiable first-century story-collector, had a story about the music of the heavens. He attributed it to Pythagoras, of course, and perhaps he was right—who knows?

> But occasionally Pythagoras draws on the theory of music, and designates as a tone the amount by which the moon is distant from the earth; its distance to Mercury half as much, and from this to Venus the same amount; from Venus to the sun one and a half tones; from the sun to

---

[9] Von Jan, "Die Harmonie der Sphären."

Mars a tone, that is, the same as from earth to the moon; from Mars to Jupiter half a tone, and from it to Saturn half a tone; thence a tone and a half to the zodiac: and so the seven tones produce the harmony called a diapason, i.e., a universal harmony.[10]

In other words, the sum of eight intervals between the nine bodies— Earth, seven planets (including the Sun and Moon), and zodiac— come to seven whole tones, an octave, or *diapason* in transliterated Greek. An elementary problem with this arrangement is that an octave equals six whole tones, not seven. Pliny has made one or more of the intervals too large.

The specific interval in error can be identified with the aid of a passage by the third-century writer Censorinus. In chapter 13 of his compendium *De die natali*, written A.D. 238, Censorinus cites Pythagoras for the same sequence of musical intervals between the planets, except for the last one:

Accordingly, Pythagoras estimated that from the Earth to the Moon there are about 126,000 *stadia*, and that makes the interval of a tone; from the Moon to the planet Mercury, which is called *stilbon*, half as much, or a half-tone; hence to *phosphoron*, which is the planet Venus, about the same amount, that is, another half-tone; and thence to the Sun three times as much, thus a tone and a half. And so the Sun is three tones and a half distant from the Earth, which is called a fifth; from the moon two tones and a half, which is a fourth. Now from the Sun to Mars, whose name is *pyrois*, is the same interval as from the Earth to the Moon, and that is a tone; hence to Jupiter, which is called *phaethon*, half of that, which makes a half-tone; the same amount from Jupiter to Saturn, for which the name is *phaenon*, that is another half-tone; thence to the highest heaven, where the signs [of the zodiac] are, another half-tone. And so the distance from the highest heaven to the Sun is a fourth, that is, two tones and a half; to the Earth, the total from the same heaven is six tones, which is called the harmony of an octave.[11]

Evidently it was the final interval between Saturn and the zodiac that Pliny exaggerated, due to a corrupt source, a lack of understanding, or likely both. At any rate there was a tradition, very old and possibly due to Pythagoras himself, that used musical intervals as an analogy (or

[10] Pliny, *Natural History*, 2.20. I have translated from the Latin of the Loeb edition, vol. 1, pp. 226–28.

[11] Censorinus, *De die natali* (Leipzig, 1867), pp. 22–24.

even a measure) for the distances among the heavenly spheres. Censorinus even gave the conversion factor: a tone was 126,000 *stadia*, and a *stadium*, he specified, was in this context to be taken as used in Italy, equivalent to 625 paces.

This same "Pythagorean" sequence of musical intervals turned up later, slightly garbled, in the encyclopedic allegory *De nuptiis Philologiae et Mercurii* written early in the fifth century by Martianus Capella. Like Pliny, Martianus Capella placed a tone and a half between Saturn's sphere and the stars; unlike Pliny, he knew that six tones made an octave, and blithely added wrong to get the correct sum.[12] His work was widely read during the middle ages, and did its part to maintain the association of musical pitches with the planets.

The accounts described here deal only with the intervals between the planets. Ancient attempts to link individual planets with specific musical notes are terse and not fully understood. According to von Jan, such associations probably passed through two stages in what he called the first system of planetary scales. The first stage, which is hypothetical since it is not described in any surviving text, placed the Sun immediately above the Moon, where Plato had put it in *Timaeus*. By the time of Pliny (first century A.D., discussed previously), and certainly by the time of Nicomachus (early second century), who first recorded the notes of all the planets, the Sun had been moved to the middle of the planets, as shown in table 2.2. Nicomachus therefore assigned to it the important note *mese* (literally "middle"), in accordance with his general attempt to derive the names of the notes from their associated planets.

In this first system the note sequence or "scale" consists simply of two adjacent tetrachords. A tetrachord, the building block of most Greek music theory, is a sequence of four notes. The two outer notes always span the interval of a fourth and are known as "fixed" or "standing" notes, whereas the two inner notes, known as "movable," can be placed at different positions to make tetrachords of the diatonic, chromatic, and enharmonic genera.[13] Two adjacent tetrachords can either be conjunct, sharing a note, or disjunct, separated by a whole tone.

---

[12] Martianus Capella, *De nuptiis Philologiae et Mercurii* (Leipzig, 1925), book 2, pp. 169–99.

[13] On this and related questions see Barbera's article "Greece," in Randel, ed., *New Harvard Dictionary of Music.* On Greek music theory generally, the texts translated and annotated by A. Barker in *Greek Musical Writings II* (Cambridge, 1989) are extremely useful.

TABLE 2.2.
Early Planetary Scale

1	2	3
Note		Scale of Nicomachus
Nete	d'	Moon
Paranete	c'	Venus
Paramese	b♭'	Mercury
Mese	a	Sun
Hypermese	g	Mars
Parhypate	f	Jupiter
Hypate	e	Saturn

The tetrachords in table 2.2[14] are diatonic (since the ascending intervals are half-tone, tone, and tone) and conjunct (since the note mese is shared).

The modern notes shown in column 2 for the Greek note-names in column 1 are in a sense arbitrary, since the exact pitch to which Greek notes were tuned is unknown. I have adopted the common convention of placing mese at a. For convenience later on, I shall employ Kepler's usual convention for distinguishing the octaves. Notes up to and including A in the lowest space of the bass clef are written with capital letters; the next octave, from b♭ within the bass clef to a on the top line of the bass clef, is written in lowercase; and subsequent octaves are distinguished by repeated lowercase letters or, more conveniently, by one or more prime symbols, as a', a", a''', and so forth.

Jan's first system is characterized by the assignment of the lowest note, called *hypate*, to Saturn, the highest planet. The concepts of "low" and "high" were not universally used by Greek authors in referring to pitch, but when they were so used, it was in the sense opposite to the one with which we are familiar. Hypate, the name of the lowest note in this seven-note scale, literally means "highest." (*Nete* means

---

[14] Von Jan, "Die Harmonie der Sphären," pp. 18–19.

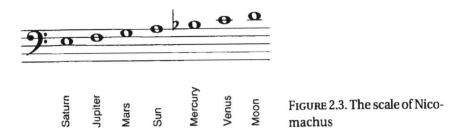

| Saturn | Jupiter | Mars | Sun | Mercury | Venus | Moon |

FIGURE 2.3. The scale of Nicomachus

"last.") Various explanations have been advanced for this convention, deriving it from such things as the positions of the strings or finger-holes on musical instruments or the heights of vertical strings needed to sound the notes.[15] At any rate, it was not paradoxical to assign hypate to the highest planet.

Figure 2.3 shows the planetary scale of Nicomachus in musical notation. The positioning of Saturn at the lowest note, here hypate, would not be seen again until Kepler inverted the music of the spheres to heliocentric form, fifteen hundred years later.

Greater and Lesser Perfect Systems

A third class of planetary scales found by von Jan was built not on the tones of a simple scale but rather on the fixed notes of a variant form of the Greater Perfect System. The Greater Perfect System (GPS) was one of the most important harmonic constructions in Greek music theory. It covered two octaves with four tetrachords, as shown in table 2.3.[16] The interior notes shown in the table for these tetrachords represent the diatonic genus of harmony.

The lower two tetrachords, hypaton and meson ("highest," i.e., lowest; and "middle"), were conjunct, sharing the note e. Separated by a tone from tetrachord meson was the tetrachord diezeugmenon ("disjunct"). Above this and joined to it, sharing the note e', was the fourth and final tetrachord, hyperbolaeon ("extreme"). An extra note, called proslambanomenos, was attached below the tetrachord hypaton to complete the double octave of the GPS.

The fixed notes of the GPS were those at the bottom and top of each tetrachord: hypate hypaton, hypate meson, mese, paramese, nete diezeugmenon, and nete hyperbolaeon. Between the fixed notes of each

[15] Barker, *Greek Musical Writings II*, p. 251–52, n. 20; Barbera, "Greece," p. 349.
[16] Barbera, "Greece," p. 349.

TABLE 2.3.
The Greater Perfect System

1	*2*	*3*
Note		*Tetrachord*
Nete hyperbolaeon	a'	
Paranete hyperbolaeon	g'	Tetrachord
Trite hyperbolaeon	f'	hyperbolaeon
Nete diezeugmenon	e'	
Paranete diezeugmenon	d'	Tetrachord
Trite diezeugmenon	c'	diezeugmenon
Paramese	b'	
Mese	a	
Lichanos meson	g	Tetrachord
Parhypate meson	f	meson
Hypate meson	e	
Lichanos hypaton	d	Tetrachord
Parhypate hypaton	c	hypaton
Hypate hypaton	b	
Proslambanomenos	A	

tetrachord were the so-called movable notes, which established the genus of the tetrachord as either diatonic, enharmonic, or chromatic. (The notes shown in table 2.3 are those of the diatonic genus, in which the upper two intervals of each tetrachord are whole tones and the lowest interval is the "Platonic limma" of 256:243.) The note proslambanomenos was conventionally included among the fixed notes as well, as if it were one end of another disjunct tetrachord. The tuning of the fixed notes was "fixed" because they did not depend upon the genus of the tetrachords. The fixed notes at the ends of any single tetrachord were separated by the interval of a fourth (4:3), whereas neighboring fixed notes in adjacent tetrachords either were separated

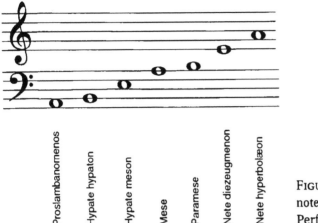

Proslambanomenos
Hypate hypaton
Hypate meson
Mese
Paramese
Nete diezeugmenon
Nete hyperbolæon

FIGURE 2.4. Fixed notes of the Greater Perfect System

by a tone (9:8), when the tetrachords were disjunct, or were the same note, when the tetrachords were conjunct. Hence the intervals separating these seven notes were identical in all variants of the GPS. Figure 2.4 shows the fixed notes of the GPS.

Closely related was the Lesser Perfect System (LPS), which was lesser because it substituted a single, conjunct tetrachord for the two upper tetrachords of the GPS. This tetrachord, called *synemenon*, spanned the fourth from a, mese, to d′, which thus acquired in this system the name *nete synemenon*. The fixed notes of the LPS are shown musically in figure 2.5.

It must be kept in mind that intervals among the notes named in the table are not identical to those among the "same" notes today, since the pitch of the system as a whole is fixed only by convention. Modern instruments, moreover, are not often tuned to precisely the Pythagorean proportions, so the intervals among notes are not exactly what would be heard on a piano, for example.

Von Jan found a third system of planetary tones, which assigned the planets to the fixed notes of the GPS, in a manuscript containing Ptolemy's musical works. This third system turns out to be one that Ptolemy had caused to be engraved, along with early parameters of his planetary models, on a stele, an upright slab of stone, set up at Canopus near Alexandria.[17]

---

[17] On the Canobic Inscription generally, see the article "The Canobic Inscription" by Hamilton, Swerdlow, and Toomer, which has overturned previous opinion and shown the Inscription to be authentic and earlier than the *Almagest*.

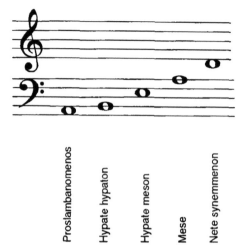

Proslambanomenos Hypate hypaton Hypate meson Mese Nete synemmenon

FIGURE 2.5. Fixed notes of the Lesser Perfect System

The Canobic Inscription

The Canobic Inscription preserved quite an unusual text. The stele itself is long gone. Thus the text, despite the seeming permanence of its original medium, is known only from manuscript copies and has been subject to the vagaries of centuries of copying and recopying. There are numerous textual problems, not all of which can be resolved. The Inscription consists of a long list of numerical parameters, starting with the obliquity of the ecliptic, continuing through the eccentricities, epicyclic radii, mean motions, and other parameters of Ptolemy's current models for the planetary motions. It is in effect a summary of the numerical facts of astronomy, as Ptolemy knew them at the time, without explanation or discussion. At the end is a table associating the planets with uninterpreted numbers. These numbers are in harmonic proportions.

Despite many attempts, the table is not fully understood. Its text is controversial to the extent that it is uncertain whether the planets have been displaced by a line, associating each planet with the wrong note. These problems are perhaps insoluble, and I shall not attempt to resolve them. Table 2.4[18] presents a possible version (Heiberg's) of the harmonic table from the end of the Canobic Inscription.[19] Column 4

[18] Heiberg, *Claudii Ptolemaei opera quae extant omnia*, pp. 154–55. I have added [Nete] before diezeugmenon. For alternatives see Halma, *Hypotheses*, pp. 57–62; and von Jan, "Die Harmonie der Sphären," pp. 29–30.

[19] The agreement of these planetary notes with those in book 3, chap. 16, of Ptolemy's *Harmonics* argues strongly in favor of this arrangement, if the text of that chapter is accepted as genuine. See Barker, *Greek Musical Writings II*, p. 390 nn. 87 and 89.

TABLE 2.4.
Planetary Harmonics from the Canobic Inscription

1 Body	2 Note	3 Number	4 Interval
Fixed stars	Mese hyperbolaeon	36	9:8 (tone) above Saturn
Saturn	Nete hyperbolaeon	32	4:3 (fourth) above Jupiter
Jupiter	[Nete] diezeugmenon	24	4:3 (fourth) above Sun 9:8 (tone) above Mars
Mars	Nete synemenon	$21\frac{1}{3}$	4:3 (fourth) above Venus & Mercury
Sun	Paramese	18	9:8 (tone) above Venus & Mercury
Venus & Mercury	Mese	16	4:3 (fourth) above Moon
Moon	Hypate meson	12	4:3 (fourth) above fire & air
Fire, air	Hypate hypaton	9	9:8 (tone) above water & Earth
Water, Earth	Proslambanomenos	8	

was not a part of the inscription at Canopus; I have added it to clarify the numbers in column 3.

The notes named in column 2 include all the fixed notes in both the GPS and the LPS, plus an extra note, called mese hyperbolaeon, placed a tone higher than nete hyperbolaeon.[20] The numbers in column 3 are simply numbers. There is no indication that they were thought to represent the planets' distances, speeds, powers, or anything else in particular. Their proportions (given in column 4) are precisely those required by the intervals between the notes in column 2. The first few ratios at the bottom, starting from proslambanomenos, are respectively a tone, a fourth, a fourth, and a tone. The next note, nete synemenon (whose nonintegral number seems at first to stand out from

[20] In book 2, chap. 6 of the *Harmonics*, Ptolemy showed the advantages for modulation of this mixed system, which he called the "complete system," combining the GPS and the LPS.

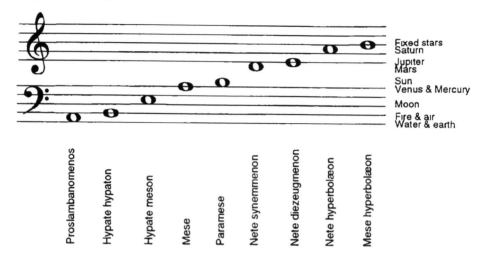

FIGURE 2.6. Ptolemy's planetary music

the rest), is in fact precisely a fourth above mese and precisely a tone below the next note, nete diezeugmenon. The latter is also a fourth higher than paramese. The last two intervals at the top are respectively a fourth and a tone. The numbers represent the notes in figures 2.4 and 2.5, with one extra note, mese hyperbolaeon, at the top. Thus Ptolemy's planetary notes were probably those shown in figure 2.6.

Whatever the rationale for associating a particular component of the heavens with its note, Ptolemy's intent was clearly to assign each of the individual bodies to one of the fixed notes of his complete system. The harmonic content of the table is much more evident than the astronomical significance. To make sure the arithmetic and harmonic elegance latent in the nine numbers of column 3 were not missed, the Canobic Inscription concluded with a tabulation of the forms of arithmetic and musical harmony in the numbers: five arithmetic means, six geometric means, five harmonic means; five fourths, four fifths, five octaves, two octave-plus-fifths, two double octaves, three tones (there are actually four tones). As a service for the skeptical reader we enumerate all these in tables 2.5 and 2.6.

Ptolemy's Harmonics

The numbers and words carved into the stele at Canopus were by no means Ptolemy's only legacy on the harmonies of the spheres. His treatise on *Harmonics* is one of the most carefully argued and influen-

Table 2.5.
Means from Canobic Inscription

1 Five arithmetic means	2 Six geometric means	3 Five harmonic means
8:12:16	8:16:32	8:12:24
8:16:24	9:12:16	9:12:18
12:18:24	9:18:36	12:16:24
12:24:36	12:16:21$\frac{1}{3}$	12:18:36
16:24:32	16:24:36	18:24:36
	18:24:32	

tial works on music theory to survive from antiquity.[21] Its astronomical content, however, is disappointing and seems not to have been a serious concern. (In this respect the book resembles Kepler's youthful *Mysterium cosmographicum*, which is similarly enthusiastic about a Truth glimpsed in the heavens.) Yet the theoretical development of harmonics is rigorous. Astronomical matters are stated confidently, even where they lack the sophistication one would expect from the writer of the *Almagest*. With the *Harmonics* the harmony of the spheres became less a matter of oracular revelation, acquiring some measure of support from a more general mathematical theory.

In harmonics as in everything else, Ptolemy went his own way. Nevertheless his views on musical harmony aligned him in large part with the Pythagorean school. Like the Pythagoreans, he thought that harmony in music was based on the mathematical proportions that arose out of the physical phenomena of sound. The school of Aristoxenus, on the other hand, treated musical pitches as absolute and independent of any other science. Since for Ptolemy the mathematical principles underlying musical harmony were also at work throughout the universe, it was natural that he thought he could find them expressed in the heavens.

Ptolemy's own theories were characterized, in harmonics as in other

[21] Ptolemy's *Harmonics* has been translated by A. Barker in *Greek Musical Writings II: Harmonic and Acoustic Theory*, chap. 11.

TABLE 2.6.
Harmonic Intervals from the Canobic Inscription

1 Five fourths (4:3)	2 Four fifths (3:2)	3 Five octaves (2:1)	4 Two octave + fifths (3:1)	5 Two double octaves (4:1)	6 Four tones (9:8)
12:9	12:8	16:8	24:8	32:8	9:8
16:12	18:12	18:9	36:12	36:9	18:16
21⅓:16	24:16	24:12			24:21⅓
24:18	36:24	32:16			36:32
32:24		36:18			

fields, by a respect for observed phenomena. In harmonic theory, this meant that the ear's judgments about what sounded pleasing were an important, and perhaps the final, criterion for choosing among theoretical possibilities. The goal was not to determine rationally the most elegant or appealing theory of harmony but to use reason and experience jointly. Theories were submitted to empirical test, and, conversely, the ear was trained to perceive distinctions proposed by reason. Ptolemy yielded to the judgments of the ear but recognized that for subtle judgments the ear required theoretically informed assistance. He devoted a great deal of attention to the proper construction of musical instruments that were scientific, in that they were designed for the specific purpose of helping a listener discriminate between different harmonic theories.

The *Harmonics* contains three books. The first two, and the very beginning of the third, treat the fundamentals of the subject, discussing such concepts as the GPS, the LPS, Ptolemy's combination of the two, and the proportions among the movable notes in all the genera. Starting with the third chapter of book 3, Ptolemy takes up the relation of harmony to other entities, in particular to human souls and to the movements of heavenly bodies.

The third chapter presents an impressively reasoned justification for applying the theory of harmony to much broader questions than those of music alone. The power (*dynamis*) of harmony, Ptolemy argues, is one of Reason rather than of God or of Nature. Harmony is not that which is eternally good, nor is it a thing at all; rather it is an active principle that infuses excellence into things. It is, in part, the form of Reason that brings order to what is heard. More generally, though, this type of power relies on both the two highest senses, sight and hearing, the senses capable of judging beauty as well as pleasure. Each of these two senses is responsible for, and in a way gives rise to, a particular science. The science that depends only upon sight is astronomy, for the heavenly bodies are perceived in no other way. Analogously, the science that depends only upon hearing is harmonics. Thus the sciences of astronomy and harmonics are related in their very foundations. Each attempts to systematize information received from one of the senses capable of perceiving beauty.

Harmony is revealed most perfectly in the movements of those things that are most rational. Among mortal things the most rational are human souls; among eternal and divine things the most rational are the heavenly bodies. Here is the fundamental rationale for much of astrology. Looking first at the soul, Ptolemy proposes detailed analogies of its parts, according to any of several extant classifications, with the different intervals used in harmony and with the three genera recognized in ancient harmonics: the diatonic, chromatic, and enharmonic. Similarly, he relates harmonic modulations to changes in souls. These analogies form a basis for the astrological use of harmonic principles.

From chapter 8 through the end of book 3, Ptolemy is concerned with analogies between concepts used in musical harmony and particular motions in the heavens. He first compares the Greater Perfect System with the zodiac, imagining it to be wrapped around the ecliptic into a circle. The central note mese is at one of the equinoctial points, and the extreme notes proslambanomenos and nete hyperbolaeon are joined at the other. This arrangement aligns the interval of an octave with one-half of the circle of the zodiac, and thus with the astrological aspect of opposition, in which two planets face each other from diametrically opposite positions in the zodiac. The other consonant intervals correspond similarly to aspects in the zodiac. The fifth arises from a proportion of 3:2 and corresponds to the trine aspect (120°), because two-thirds of the way around the zodiac from any point is a position in trine aspect to that point. Similarly the fourth, 4:3, corresponds to quartile aspect (90°).

Beginning in chapter 10 of book 3, Ptolemy compares the various motions of planets to the kinds of "movement" found in music. Motion *kata mēkos*, "in length," which probably means motion along the celestial equator, is like rising and falling pitch. Low pitches are associated with rising and setting points, and high pitches with the mid-heaven, farthest from the horizon. Motion *kata bathos*, "in depth," between positions near the Earth and positions highest above the Earth, is like the transitions among the three genera of harmonic theory. The lowest position corresponds to the diatonic genus, the middle position to the chromatic genus, and the highest position to the enharmonic genus. Finally, motion *kata platos*, "in breadth" or away from the celestial equator, is analogous to modulations among the *tonoi*, or modes, used in music. The sounds described in all this seem to be essentially a low tone as the planet rises in the east, ascending to a high tone when the planet culminates, and then descending again until the planet sets in the west. Configurations of planets relative to the sun are analogous to the different arrangements of tetrachords and disjunctive tones within the GPS.

At the end of book 3 Ptolemy suggests, in chapters whose contents were mostly lost by the middle ages, that the sizes and motions of the planetary spheres are in proportions determined by the same fundamental laws that rule the science of harmonics.[22] The title of the fourteenth chapter asks in which numbers the fixed notes of the complete system may be compared with the planetary spheres. This is obviously reminiscent of the lists on the Canobic Inscription, discussed above, and it is safe to assume that the lost text of the chapter included that material. Perhaps it explained what meaning, if any, Ptolemy thought the planets' numbers to have aside from their harmonic ratios. If so, its loss is very serious indeed, for as matters stand we have essentially no idea of the intended astronomical significance of those numbers.

Chapter 15 of book 3 dealt with the ratios of the movements of the various planets; aside from its title is totally lost. Presumably it contained a quantitative account of the three movements (in length, depth, and breadth) discussed in chapter 10. Chapter 16, by its title, discussed analogies between the relations among the planets and those among the notes. If the text located by Nikephoros Gregoras (see

---

[22] Barker, *Greek Musical Writings II*, pp. 371–91. The text now associated with chaps. 14 and 15 was written by Nikephoros Gregoras, a fourteenth-century Byzantine scholar, to fill in the gap in the manuscripts. He found a text that is probably Ptolemy's original chap. 16 misplaced earlier in one of the manuscripts.

n. 22) is authentic, as it seems to be, the chapter relates the musical harmonies to astrological characteristics of the planets. It discusses the propitious or malevolent characters of the planets as they relate to their positions on the fixed notes of the perfect system. This provides a broad and deep theoretical foundation for some doctrines, at least, of Ptolemy's astrology. It may even be that Ptolemy determined the planets' musical notes by studying the astrological relations among them. But we do not know.

PTOLEMY'S HARMONICS AND KEPLER

It must be understood that Ptolemy merely sketched these analogies, providing little detail. His usage of many musical concepts, including genus and especially tonus, is not fully understood even today.[23] What is clear today—and was clear to Kepler at the beginning of the seventeenth century—is that Ptolemy thought that orderly motion, in the heavens as in music, followed only certain kinds of patterns, so that study of the patterns in one field could in theory elucidate those in the other. Rational motion obeyed the same laws everywhere, in the celestial spheres as in the strings of the *lyra*, not for any mystical reason but precisely because those were the laws of rational motion. Astrological influences, which were notoriously difficult to analyze because of the great number of interrelating causes, might profitably be studied by considering the effects of various kinds of music on the human soul. The motions of the planetary spheres could similarly be understood more deeply through an awareness of the principles they shared with musical harmony. In Ptolemy's *Harmonics*, connections such as these were assumed to be rational, although they were not assumed to be understood—yet—in detail.

It was these implications of Ptolemy's harmonic theories that aroused an intense interest in the young Johannes Kepler. When the Bavarian Chancellor Herwart von Hohenburg mentioned to Kepler some of Ptolemy's opinions from his *Harmonics*, in a letter of 1599, the young astronomer, only twenty-seven and already interested in harmonies that he had found manifested in the heavens, eagerly requested that Herwart send him the book. (It turned out to be a bad Latin translation.)[24] By the time he finally received a manuscript of the

---

[23] Barbera, "Greece," pp. 346–51.

[24] Herwart to Kepler, 29 August 1599; Kepler to Herwart, 14 September 1599, nos. 133 and 134 in *G.W.* 14.59–76.

Harmonics in its original Greek, in 1607, he had drafted a book of his own on the subject, apparently an early version of what later became the *Harmonice mundi.* A decade later he took up Ptolemy's *Harmonics* in earnest, preparing a translation of that work to accompany his own, which he was finally preparing to publish. He never did get the translation into print, and it remained among his papers at his death.

There can be little doubt that Kepler's painstaking investigations into the sublime harmonies of the planets were inspired by the belief that the greatest astronomer of antiquity had attempted to pursue a similar research program. He regarded Ptolemy's work in harmonics in much the same way as his work in astronomy: he regarded both as brilliant, even inspired, attempts to answer questions of fundamental importance. At the same time, however, he recognized that Ptolemy was not and could not have been successful, because the state of knowledge in his time had been inadequate for him to grasp the truth that Kepler perceived. On the basis of this essentially historical view of Ptolemy, Kepler reconstructed, as we shall see, the missing last three chapters of the *Harmonics.* We shall devote chapter 7 to Kepler's reconstruction and analysis of his predecessor's work.

Cicero and Macrobius

Macrobius, who was active around A.D. 400, was one of the "encyclopedists," a miscellaneous group of writers (including Pliny, Martianus Capella, Boethius, and Isadore of Seville) who transmitted a broad but superficial knowledge of the literature of antiquity to the middle ages. He is best known for his commentary on the Dream of Scipio, a short passage from the final section of Cicero's *De re publica.* The text of *De re publica* was lost in the middle ages, aside from a few short passages and the complete Dream of Scipio, which survived by being copied into manuscripts to accompany Macrobius's commentary on it.[25]

Cicero had discussed in a few eloquent pages themes of duty, astronomy, geography, and the insignificance of human glory. Macrobius's commentary ran sixteen or more times the length of the original, taking nearly every concept mentioned in the original as an invitation to recount the opinions of other authors on the subject. Containing little or no original thought, it was a clearly written compendium that preserved excerpts of many works that would otherwise have been

---

[25] Stahl, p. 10.

lost. The theories of the neo-Platonists, which Macrobius probably had from Porphyry, figured prominently in the discussion.[26]

One of the neo-Platonic fables that Macrobius preserved, in Cicero's words, is literally that of the harmony of the spheres: " 'What is this great and pleasing sound that fills my ears?' I asked. 'That,' replied my grandfather, 'is a concord of tones separated by unequal but nevertheless carefully proportioned intervals, caused by the rapid motion of the spheres themselves. The high and low tones blended together produce different harmonies. . . . [The] outermost sphere, the star-bearer, with its swifter motion, gives forth a higher-pitched tone, whereas the lunar sphere, the lowest, has the deepest tone.' "[27]

The characterization of the motions in the final sentence is correct from a geocentric point of view. The sphere of the stars is swiftest because it rotates daily, from east to west. The lower spheres lag behind, seeming to move from west to east with respect to the sphere of the stars. (They still rotate almost daily about the Earth, in absolute terms.) The sphere of the Moon lags by one rotation per month, being slower thus by about one part in thirty.

In his commentary, Macrobius remarks that the motions of such vast bodies as the heavenly spheres surely produce noise. This noise would be either sweet or harsh, depending upon whether the strokes that caused it were "in keeping with certain numerical relations" or not. But motions in the heavens are certainly orderly and follow divine law. Hence the noise of those motions is surely melodious.

Credit for first noticing the numerical relationships involved in harmonious sounds is due to—who else?—Pythagoras. Macrobius recites the old story about Pythagoras noticing a harmony among the clangor of blacksmiths' hammers, investigating the weights of the hammers, transferring his researches to plucked strings under well-proportioned tensions, and thereby discovering the science of harmony. The crucial proportions were of course the consonances found among the numbers 1, 2, 3, and 4: the fourth, fifth, octave, octave-plus-fifth, and double octave.[28]

Macrobius digresses next on the three-dimensionality of bodies, from which he somehow arrives at the Pythagorean or Platonic sequence of doubles and triples 1, 2, 3, 4, 9, 8, 27. After admiring this

[26] Ibid., pp. 3–12.
[27] Ibid., p. 73.
[28] Ibid., pp. 185–88.

TABLE 2.7.
"Platonic" Planetary Distances According to Macrobius

1	2	3
Distance to Sun is	2	times distance to Moon
Distance to Venus is	3	times distance to Sun
Distance to Mercury is	4	times distance to Venus
Distance to Mars is	9	times distance to Mercury
Distance to Jupiter is	8	times distance to Mars
Distance to Saturn is	27	times distance to Jupiter

sequence and explaining how necessary it is for a proper understanding not only of Cicero but of the motions in the heavens, he turns his attention to the Sirens sitting on each of the heavenly spheres in the myth of Er from Plato's *Republic*.[29] This leads finally to the sizes of those spheres. Macrobius reports that Archimedes calculated these sizes but that his results are rejected by the Platonists "for not keeping the intervals in the progressions of the numbers two and three." The true Platonic distances are shown in table 2.7.

One need not look far for the origin of these astronomical parameters: column 2 contains the magic sequence of doubles and triples, three of each. Macrobius, unlike Plato, uses the sequence multiplicatively, multiplying each distance by the next magic number to get the next distance. (Hence he is able to retain the "natural" order of terms, with the second square, 9, preceding the third cube, 8.) The order of the planets here, with the Sun next to the Moon, was old, as in Plato. The distances themselves were inconsistent with statements elsewhere in the *Commentary*. Macrobius is not abashed. He retreats to a conventional recital of the causes of pitch and then summarily drops the whole subject of music, since it would be pedantic rather than useful, he modestly insists, to discourse on "the nete and hypate and the other strings and to discuss the subtle points of tones and semitones."

Macrobius does not associate specific planets with specific notes of

29 Plato, *Republic*, book X, 617b; in Hamilton and Cairns, p. 841.

the scale, as had Nicomachus and Ptolemy. No doubt he would have, if he had read the works of those authors on harmony. Yet he transmitted the tenets of the harmony of the spheres and associated the fundamental harmonic number-sequence with the planetary distances. The theme was woven deeply enough into the written culture of the ancient world to survive the centuries in which that culture slowly unraveled. The idea preserved by Macrobius and the other encyclopedists—that the arrangement of the heaven was based on harmonic principles—became a standard, if little-comprehended, part of medieval accounts of the theory of music.

Boethius

The most important of the encyclopedists for later music theory was Boethius (c. 480–524), a younger contemporary of Macrobius. His writings on Aristotle's logic and on the sciences of the quadrivium (arithmetic, music, geometry, and astronomy) supplied a readable Latin compendium of ancient Greek thought. His treatise *De institutione musica*, which C. Palisca has described as "partly translation, partly paraphrase, and partly commentary on Greek treatises,"[30] was the most influential source of Pythagorean music theory throughout the middle ages.

Boethius expounded a threefold division of "music" into *mundana, humana,* and *instrumentalis.* Of these only the third corresponds to what we know today as music. *Musica mundana* is the harmony of the heavenly spheres, and *musica humana* deals with the influence of music, broadly considered, on the human soul. The very expansive conception of music embraced by these three categories provides the theoretical basis for most discussions of music into the sixteenth century and indeed underlies the astronomical and astrological systems of harmony found in books 5 and 4, respectively, of Kepler's *Harmonice mundi.* Thus for Boethius and his readers there is nothing fanciful about the music of the heavens. It is a standard part of the subject. If writers on music give less emphasis to *musica mundana et humana* than to instrumental and vocal music—as, indeed, does Boethius—they are at least aware that they are omitting a large part of their subject.

[30] Claude V. Palisca, *Humanism in Italian Renaissance Musical Thought* (New Haven: Yale University Press, 1985), p. 36.

TABLE 2.8.
Planetary Scale of Boethius

1 *Planet*	*2* *Note*
Saturn	Hypate meson
Jupiter	Parhypate [meson]
Mars	Lichanos meson
Sun	Mese
Venus	Trite synemmenon
Mercury	Paranete synemmenon
Moon	Nete [synemmenon]

After explaining which proportions corresponded to the various consonances in Pythagorean music theory, Boethius devotes a perfunctory page to the comparison of specific strings (that is, notes) with particular planets.[31] His arrangement is shown in table 2.8.

This scale is recognizably that of Nicomachus (see table 2.2), although for Mars and Venus Boethius uses the regular note-names from the Lesser Perfect System instead of paramese and the unusual hypermese employed by Nicomachus.

Boethius continues by noting blandly that Cicero, in book 6 of *De re publica*, arranged things in the opposite order, with Earth silent at the center and the planets as shown in table 2.9.

This is the order of planetary notes in Cicero's Dream of Scipio, upon which Macrobius wrote his commentary. The specific choice of notes, however, is not given by Cicero (or by Macrobius), and it may be that Boethius supplied them according to his own taste. At any rate, he does not comment at all on the discrepancy between the two systems or even explicitly state a preference between them.

Boethius, indeed, shows little interest in the details of *musica mundana*. He seems to have included the planetary scales in his musical treatise for the sake of completeness. It was not a subject about which

[31] Boethius, *De institutione musica*, vol. 1, chap. 27; p. 131 in the Marzi edition.

TABLE 2.9.
Planetary Scale of Cicero According to
Boethius

1	2
Planet	*Note*
Moon	Proslambanomenos
Mercury	Hypate hypaton
Venus	Parhypate hypaton
Sun	Lichanos hypaton
Mars	Hypate meson
Jupiter	Parhypate meson
Saturn	Mese

he had anything to say, despite the importance he accorded it as one of the three divisions of music. He repeated the planetary scales simply because they were there and because they seemed likely to be important. With this attitude he set the pattern for most references to the subject in the next few centuries.

Musica mundana in the Renaissance

The Boethian *musica mundana, musica humana, musica instrumentalis* categorization was widely accepted by most theorists in the middle ages. By the fifteenth century, however, dissenting opinions to the idea of *musica mundana* were voiced. The Florentine humanist Coluccio Salutati wrote in the 1380s a treatise *De laboribus Herculis*, never published, in which he denied that the motions of the heavens could produce sound. Within the ranks of musicians, Johannes Tinctoris (of French birth but working for Ferdinand I of Naples) picked up on Aristotle's denial that any sound arose from the motions of the heavenly spheres: "Hence I could never be persuaded that musical harmonies, which cannot be produced without sound, are made by the motion of the heavenly bodies." The real crux here was an insistence that musical harmony could not exist without physical sound. This position of Tinctoris abandoned the broad, mathematical conception of harmony found in Boethius and in the Pythagorean tradition generally.[32]

Despite such challenges, the concept of *musica mundana* remained strong in fifteenth-century Italy. Marsilio Ficino's translation of the works of Plato, and of the *Timaeus* in particular, reinvigorated the notion of celestial harmony. Franchino Gaffurio (1451–1522), the most important music theorist of his time, was thoroughly Pythagorean in his approach to the subject. He accepted as consonances only the intervals approved by Boethius, which did not include the major and minor thirds and sixths—a matter of some concern, since working musicians were using thirds and sixths as consonances. Gaffurio treated the music of the spheres in some depth, affirming that this music could be heard by truly virtuous men.[33]

Gioseffo Zarlino and Vincenzo Galilei

As Italian music theorists devoted more attention to the practical problems of composition, the traditional divisions of *musica mundana* and *musica humana* were slighted. Gioseffo Zarlino (1517–90), choirmaster at San Marco in Venice, dominated the field of music theory during the last decades of the sixteenth century. His great *Istitutioni harmoniche*, first published in 1558, was the last attempt to include all of music, as defined by Boethius, in a single treatise. It included a serious discussion of the size of the cosmos, one of the traditional questions of *musica mundana*. Unlike Gaffurio, Zarlino did accept thirds and sixths as consonances. He justified their presence by extending the fundamental Platonic and Pythagorean number sequence from 1, 2, 3, 4 to include 5 and 6 as well—6 being the first "perfect number," which equals the sum of its factors 1, 2, and 3. By allowing 5 and 6 within the fundamental sequence, Zarlino could accept thirds corresponding to proportions of 4:5 and 5:6, and the major sixth corresponding to 3:5. This extension made Pythagorean harmonic theory much more plausible as a basis for contemporary composition.

---

[32] Palisca, *Humanism*, pp. 181–85; Moyer, pp. 63–67. The quotation is from E. de Coussemaker, ed., *Oeuvres theoriques de Jean Tinctoris* (Lille, 1875), p. 200.

[33] Moyer, *Musica Scientia*, pp. 67–92; Palisca, *Humanism*, pp. 166–77.

Zarlino was particularly interested in the emotional effects of music. He argued, more thoroughly than previous scholars, that ancient music had been essentially monophonic, with only a single note sounding at a time, and that misunderstanding of this point had contributed greatly to the difficulty of understanding ancient music.[34]

Zarlino established the framework for "modern" music theory as Kepler understood it. Nevertheless, it was Vincenzo Galilei (late 1520s–91), a onetime student and frequent opponent of Zarlino, and the father of Galileo, whom Kepler cited more often by far. This was, no doubt, because he read "three-quarters" of Galilei's *Dialogo della musica antica, et della moderna* (1581) as distraction on a slow, sad journey in 1617.[35] It was fresh on his mind when he resumed work on the *Harmonice mundi*.

Galilei's treatise drew on the ideas of Girolamo Mei (1519–94), who corresponded at length with Galilei but did not publish on his own. Mei had concluded, like Zarlino, that ancient music was innocent of polyphonic harmony. He was not a practicing composer and felt no qualms about judging the musical ideals of the ancients to be superior to those of his contemporaries. For example, Mei considered polyphony a debased form of music, founded on the pleasures of sensory perception rather than the high precepts of reason. Galilei himself agreed that ancient music was superior to modern. He attacked Zarlino, and the Pythagorean tradition in general, for having oversimplified the relationship between music and their beloved numerical ratios. Instead of being the essence of music, numbers represent measurements, measurements of a host of related quantities bearing upon the production of musical notes. As Moyer has put it, Galilei's findings about the empirical basis of musical consonances "dealt the claims of Pythagorean theory to universality a blow from which it never fully recovered." In a sense, Galilei completed the process of wrenching music, in the sense we use the word, from the all-embracing *musica* of Boethius.[36]

But not for Kepler. Despite the pleasure he took in reading Galilei's *Dialogo*, Kepler steadfastly adheres to the full-blown Pythagorean conception of musical harmony. To him arguments about whether

[34] Moyer, *Musica Scientia*, pp. 202–9.

[35] Kepler to M. Wacker von Wackenfels, beginning of 1618, no. 783 in *G.W.* 17.254. This was in the midst of the witchcraft proceedings against Kepler's mother; see Caspar, *Kepler 1571–1630*, (New York, 1962), pp. 248–67.

[36] Moyer, *Musica Scientia*, pp. 225–63, 282.

actual sounds are produced by the heavenly spheres are beside the point. Ironically, Kepler agrees with Tinctoris (and Aristotle) that there is no real sound in the heavens.[37] Real sound, a matter of such importance to the music theoreticians of the Renaissance, is by no means a necessary part of what Kepler understands as the music of the heavens.

[37] "Jam soni in coelo nulli existunt, nec tam turbulentus est motus; ut ex attritu aurae coelestis eliciatur stridor." *Harmonice mundi,* book 5, chap. 4; *G.W.* 6.311.33–34.

Jofrancus Offusius:

SCIENTIFIC ASTROLOGY BASED ON HARMONY

AMONG THE MANY talented eccentrics whose pens were loosed by the turmoil of the sixteenth century was the itinerant astrologer Jofrancus Offusius, of whom remarkably little is known today. Offusius would probably be forgotten altogether were it not for a couple of cautiously favorable remarks by Tycho Brahe about his estimate of the distance to the Sun—about which more is to follow—and a couple of unfavorable remarks by Kepler about his astrological theories.[1] His background is unknown; he describes himself as a German lover of knowledge.

In 1570 Offusius published a small theoretical tract entitled *Concerning the Divine Power of Heavenly Bodies, against Demented Astrology*, in which he professed the hope that posterity might no longer "wallow in the filth" of the astrology then current. As the foundations for his reform he proposed mathematical theories, by which he hoped to make astrology truly scientific. Along the way he revealed the patterns that he had found in the distances, sizes, and eccentricities of the planets.[2]

Offusius claims to have made about twenty-seven hundred observations of the heavens during his travels, as he searched for someone who could answer his questions.[3] These were surely not observations of any great precision; but the claim exemplifies his self-proclaimed respect for astronomical observation. This respect for observation is,

---

[1] Tycho Brahe, *Opera Omnia*, ed. Dreyer (Copenhagen, 1913–29), 2.421–22; 7.328; cited by V. Thoren, *Lord of Uraniborg* (Cambridge, 1990), pp. 99, 303. Kepler, *G.W.* 1.190.21 (*De stella nova*, c.9); 15.234.123 (Kepler to Christoph Heydon, October 1605).

[2] Jofrancus Offusius, *De divina astrorum facultate in larvatam astrologiam* (Paris, 1570). We can be certain that Kepler read this book: in chap. 9 of *De stella nova* he cited Offusius disapprovingly for opinions that were advanced in chap. 16 of *De divina astrorum facultate*. Offusius also claimed to have written a *Theoria* treating the motions of the planets, and a detailed refutation of Cardano's great *De subtilitate rerum*; I have been able to learn nothing of these works.

[3] Offusius, *De divina astrorum facultate*, fol. ii^v.

in large part, the obverse of a vigorous contempt for the academic world. He says that he has found many professors of astronomy to be the type of whom Regiomontanus remarked that they observed inside their huts rather than in the great observatory of the heavens. Although they claim at first to have made numerous observations of their own, they soon retreat upon being shown a few that Offusius has made. They are, he writes scornfully, sycophants rather than philosophers, drones flying among the bees, and "demented asses worthy of the lash."[4] Needless to say, he professes nothing but respect and admiration for men who are truly learned.

ASTRONOMICAL AND ASTROLOGICAL PRINCIPLES

Offusius's cosmology is geocentric, as one expects from a sixteenth-century astrologer. He is familiar, however, with the work of Copernicus. He not only refers to the numerical parameters resulting from that work, which after all were widely circulated because they were based in part on new observations, but cites arguments from specific chapters of De revolutionibus.[5] Like many of his contemporaries who were aware of Copernicus, Offusius regards him not as one who has shattered traditional notions of Heaven and Earth but simply as a diligent astronomer who has taken the trouble to recalculate the parameters of all the motions and has supplemented the ancient observations with newer ones, including some he has made himself. The Copernican theory that the Earth moves is, to Offusius, only hypothetical. After all, the unsigned preface to De revolutionibus that concedes as much was, in the sixteenth century, generally taken to be the work of the author. (It was only Kepler who learned—and published—that Osiander and not Copernicus had written that preface.)

For Offusius the importance of Copernicus is rather that he has a greater store of observations, covering a longer period of time, than any of the astronomers who preceded him: "Copernicus, a man not inferior to the others, or at least born to a better fortune than they (for he alone used the observations of all as well as more recent ones) demonstrates that the errors of both [Mercury and Venus] are savable.

[4] Ibid. fol. iii.

[5] For example Offusius's note on fol. 3 cites book 4, chap. 19 of De rev. on the geometry of shadows in eclipses.

if they travel around a quiescent sun."[6] In making this assertion about Copernicus, Offusius is as much an instrumentalist as Osiander was in the preface to *De revolutionibus*. He reasons, explicitly, that since the authorities disagree as to the order of Mercury, Venus, and the Sun, and there is no observational way of deciding among them, he is entitled to place these planets in whatever order suits his own theory.

Offusius seems, in fact, to regard astronomy as a collection of facts—observations—organized by competing opinions. If his own assumptions agree with any of these opinions, no one can object, for that opinion might be the correct one. If numerical results that he deduces, using his own methods, are close to the consensus values, he considers his methods to be confirmed by the agreement. Holding such views on scientific methodology, he is not one to be persuaded easily that the Earth beneath his feet is spinning through space.

He is, it seems, a man who likes to speculate about great matters, who likes equally to calculate precise answers to questions, but who goes out of his way to avoid direct conflict with received wisdom—not an uncommon type then or now, and certainly not among astrologers. The received wisdom of his age ("manifest experience," he calls it, "which it is sophistical to oppose") is that the effects of planetary configurations on terrestrial affairs are due to the "rays" of the planets. The rays of the Sun make things hot and dry, while those of the Moon moisten things and gradually cool them. Saturn makes things cold and dry. Jupiter brings a very mild heat and moisture. The rays of Mars burn and desiccate, while those of Venus bestow moisture and a weak heat. The only effect of Mercury's rays is to dry out whatever they touch.[7]

Offusius provides, as a starting point for his calculations, the proportions among the strengths of the four qualities received from each of the planets. He admits that the actual demonstration of these proportions is difficult if not impossible, but he thinks that by using conjectures agreeable to reason and experience he has attained a set of proportions both pleasing and useful. Each of the four qualities is associated with one of the five Platonic polyhedra. Table 3.1[8] shows how all this comes out numerically.

---

[6] Offusius, *De divina astrorum facultate*, fol. 6[r].

[7] Ibid., fol. 1[v]. The "principles of the qualities" were not the same as the light by which we see the planet, but they were transmitted along with that light. Fol. 19[r].

[8] Ibid., fol. 4[v]. The missing polyhedron, the dodecahedron, corresponded to the sphere of the fixed stars, and to the fifth essence.

TABLE 3.1.
Strengths of Unmodified Qualities

1 Body	2 Heat (pyramid)	3 Moisture (icosahedron)	4 Cold (octahedron)	5 Dryness (cube)
Sun	27 or 3^3			49 or 7^2
Moon		100 or 10^2	$6\frac{303}{512}$ or $\left(\frac{15}{8}\right)^3$	
Saturn			$107\frac{11}{64}$ or $\left(\frac{19}{4}\right)^3$	$12\frac{1}{4}$ or $\left(\frac{7}{2}\right)^2$
Jupiter	$\frac{1}{8}$ or $\left(\frac{1}{2}\right)^3$	16 or 4^2		
Mars	1 or 1^3			$12\frac{1}{4}$ or $\left(\frac{7}{2}\right)^2$
Mercury				$11\frac{1}{4}$
Venus	$\frac{1}{8}$ or $\left(\frac{1}{2}\right)^3$	$17\frac{121}{512}$		
Total	$28\frac{1}{4}$	$133\frac{121}{512}$	$113\frac{391}{512}$	$84\frac{3}{4}$

All the numbers associated with heat and cold, which he termed the active qualities, are cubic numbers of a sort, as shown. Most of the numbers associated with the passive qualities of moisture and dryness, excepting only the dryness of Mercury and the moisture of Venus, are square numbers.

The total amounts of each quality, in the last row of the table, are in the same proportion as the volumes of the respective polyhedra, or so Offusius claims. I frankly do not know what he means by this. At any rate, the overall total, combining the unmodified strengths of all the qualities, is exactly 360—a beautiful number whose presence here he points out as indicating the work of the Framer of Nature.[9] One supposes that Offusius built in that beautiful number himself, by calculating one of the unconstrained strengths as 360 minus the sum of the rest.

The force of these rays was, of course, modified by circumstances; and here, Offusius thinks, astrologers have been insufficiently diligent in applying mathematics to their analysis. Among the circumstances with plausible and well-attested effects are conjunctions and opposi-

---

[9] Ibid., fols. 3ᵛ–5ʳ. He justified the paltry portion of heat emitted by the planets on the grounds of that quality's ability to consume moisture furtively.

tions of different planets, the length of any one planet's visibility above the horizon, and the distance between Earth and the planet. He first takes up the effect of the changing distance from a planet to Earth.

PLANETARY DISTANCES

Offusius begins with the nearest planet, the Moon. It is, he says, undisputed or nearly so among astronomers that the Moon at mean distance is about $30\frac{1}{4}$ terrestrial diameters distant from Earth. His own calculations, he reports, give $30\frac{3}{8}$ diameters for this distance. These estimates are commendably up-to-date. Ptolemy's lunar theory had resulted in a reasonably accurate distance at apogee, but had brought the Moon much too close during much of its path. The great astronomers of the Islamic world had followed Ptolemy's lead in this, with some uneasiness at the variation that the model implied in the apparent size of the Moon. Copernicus, however, had improved the model to give a mean distance of 60;18 terrestrial radii, or 30;9 terrestrial diameters.[10] (These numbers are in sexagesimal—base 60—notation, as was standard in technical astronomy. The semicolon separates the integral portion from the sexagesimal fraction. Thus 30;9 is $30\frac{9}{60}$, slightly less than the $30\frac{1}{4}$ cited by Offusius as the consensus.) The discrepancies are small, and he asks that the reader allow him to use his own value, "for our entire purpose is a harmony of the Motions, and then of the Effects."

The distance of the Sun is agreed upon, he further insists, as about 576 terrestrial diameters, a very pleasing number because it is equal to 24 squared: "Nor should anybody object that Ptolemy (as shown in the work of great construction [i.e., the Almagest]) found 580, the philosopher Aratus 555, more recent writers [*recentiores*, meaning Copernicus] 571, and I myself writing against Cardano's *de Subtilitate* demonstrated 579. All of these are less than exact, and close to the truth."[11]

The experts are not sure of the exact distance, in other words, so Offusius feels himself justified in using a number that pleases him, so long as it is close to the expert opinions. (The number pleased Tycho

[10] A. van Helden, *Measuring the Universe* (Chicago, 1985), chaps. 3 and 4. Copernicus, *De revolutionibus*, book 4, chap. 9, on which see also N. M. Swerdlow and O. Neugebauer, *Mathematical Astronomy in Copernicus's De Revolutionibus* (Berlin and New York, 1984), vol. 1, pp. 240–41.

[11] Offusius, *De divina astrorum facultate*, fol. 3r–3v.

TABLE 3.2.
Planetary Distances According to
Offusius

1	2
	Mean distance
Planet	(terrestrial diameters)
Moon	$30\frac{3}{8}$
Venus	81
Mercury	216
Sun	576
Mars	1,536
Jupiter	4,096
Saturn	$10,922\frac{2}{3}$

also, who noted it with approval.) Unlike the lunar distance, this "generally agreed" solar distance is seriously off the mark, being too small by a factor of 20.[12]

From the Sun, Offusius jumps to Saturn, citing the opinion of an unspecified expert—actually, the ninth-century Arabic astronomer al-Farghānī—that its distance is 10,055 terrestrial diameters, and his own preference for exactly $10,922\frac{2}{3}$ t.d. Then, finally, he explains where his remarkably precise distances were coming from: he simply postulates that the mean distance of each planet from Earth is $\frac{8}{3}$ as great as that of the next closer planet, as shown in table 3.2.

The Sun is, according to the experts, about nineteen times as distant as the Moon; and Saturn is, according to other theories of the experts, more distant than the Sun in about the same proportion. With two planets to be inserted in each of these intervals, Offusius has simply divided the intervals into thirds. The cube root of 19 is about $\frac{8}{3}$, so by assuming that this ratio holds between each pair of adjacent planets, he has fixed all the mean distances at a single stroke. His starting point

[12] See Neugebauer, *A History of Ancient Mathematical Astronomy* (Berlin and New York, 1975), 1. 110–12 for discussion of Ptolemy's error in the solar distance, and its consequences. For Tycho's opinion see the *Progymnasmata* (1602), p. 472.

TABLE 3.3.
Eccentricities According to Offusius

1	2	3
	Difference between apogeal and perigeal distances (terrestrial	
Planet	diameters)	Square root
Moon	9	3
Venus	64	8
Mercury	144	12
Sun	49	7
Mars	1,369	37
Jupiter	1,296	36
Saturn	1,681	41

for the whole system must have been the solar distance of 576 terrestrial diameters, a perfect square and a number whose beauty impressed him greatly.

Taken as a group these distances bore no very close resemblance to anybody else's; but in geocentric astronomy there was no way to calculate the distances without employing doubtful assumptions. As Offusius writes, "If I also could save the appearances by means of these intervals (I do not say as easily as others, but rather more easily, or rather more truly), who will deny them to me?"[13]

Perhaps the most curious thing about the distances proposed by Offusius is his placement of Venus inside Mercury. The relative positions of these two planets and the Sun had long been uncertain, since all three revolve around the Earth at an average rate of once per year. Offusius cites the differing opinions of Plato, al-Bitruji, Ptolemy ("and after him almost all the schools of later mathematicians"), and Copernicus but holds to his own order. He can, after all, prescribe distances agreeing with observation according to any arrangement of Mercury, Venus, and Sun; and his own arrangement, as shown in table 3.3,

[13] Offusius, De divina astrorum facultate, fol. 6ʳ.

53

TABLE 3.4.
Greatest and Least Distances According to Offusius

1 Planet	2 Mean distance	3 Minimum distance	4 Maximum distance
Moon	$30\frac{3}{8}$	$25\frac{7}{8}$	$34\frac{7}{8}$
Venus	81	49	113
Mercury	216	144	288
Sun	576	$551\frac{1}{2}$	$600\frac{1}{2}$
Mars	1,536	$851\frac{1}{2}$	$2,220\frac{1}{2}$
Jupiter	4,096	3,448	4,744
Saturn	$10,922\frac{2}{3}$	$10,082\frac{1}{6}$	$11,763\frac{1}{6}$

supports another theory of his that the eccentricity of each planet, measured in terrestrial diameters, is a perfect square.[14]

Moreover, the sum of the square roots is 144, itself a perfect square. In fact, it is the square pertaining to Mercury in this very table. This is entirely appropriate, since astrologically Mercury is androgynous. It is the planet that "accommodates itself to all."[15]

From the last two tables it is trivial to calculate and display the greatest and least distances of the planets, shown in table 3.4.

Effects of the Planetary Distances

The reason that Offusius goes into such detail about planetary distances in a book about the foundations of astrology is to explain how and why the practicing astrologer should adjust planetary influences (table 3.1) for the distances of each planet at the moment of interest. The quantities given in that table are valid only at the planet's mean distance. At lesser distances they must be augmented according to an elaborate computation, and at greater distances they must be reduced according to a different elaborate computation.

A planet's influence gains strength as it approaches the Earth in a

[14] Ibid., fols. 6ᵛ–7ʳ.
[15] Ibid.

proportion given by certain rules. If a planet should approach Earth by exactly a quarter of its mean distance (a *quadrans*), its qualities would be doubled in strength. The increment gained by the approach of a quadrant is thus equal to the mean strength. In the general case, the increment is calculated using one of two formulas, depending on both the amount of approach and the quality being estimated. At approaches greater than a quadrant, the proportion varies cubically (for the quality of heat) or quadratically (for the other three qualities). At approaches of less than a quadrant the variation is quadratic for heat and cubic for the other qualities—"for this reason, that in fractional amounts the square is greater than the cube, while the opposite is true for amounts greater than one."[16] Thus if a planet could approach by two quadrants, to half its mean distance, the effect of its heating faculty would be eight times normal, and the effect of the other faculties four times normal. No planet actually approaches Earth so closely at perigee, according to table 3.4.

When a planet is farther than its mean distance, the same scheme determines the decrements in its influence—but here a problem arises. It was possible to add an increment greater than a planet's mean effusion when it approached Earth beyond a quadrant. It would be another matter altogether to subtract from the mean effusion an amount greater than itself, leaving less than nothing, when a planet withdrew more than a quadrant beyond its mean distance. (There can be no need to interpret "less than zero" of a quality, since opposing qualities—heat and cold, moisture and dryness—are already included among the effusions.)

Offusius dodges this problem, without acknowledging it, by declaring that decrements in the strength of heat are one-eighth the corresponding increments, and decrements in the other qualities one-fourth the increments. With these ad hoc adjustments the effusion of any quality would only drop to zero if it moved two quadrants beyond its mean distance—but as noted above, no planet's path is sufficiently eccentric for that to happen.

As an example of the calculations required by this theory, Offusius considers the Sun at perigee. It has approached Earth by $24\frac{1}{2}$ terrestrial

---

[16] Ibid., fols. 7ᵛ–8ʳ. The effect of splicing together quadratic and cubic functions in this way was to emphasize the effects of the qualities other than heat at approaches of about a quadrant and to emphasize the effect of heat at approaches noticeably less than or greater than a quadrant. Offusius seems to have thought that he was honoring the quality of heat by giving it always the greater of the two functions.

diameters (from table 3.3). A quadrant of its mean distance is 144 diameters, so the perigeal Sun has approached by slightly more than $\frac{1}{8}$ of a quadrant. Its heating faculty varies as the square of distance for approaches less than a quadrant, so the increment to that faculty is about $\frac{1}{36}$ of the faculty's mean strength. This mean strength is 27 (from table 3.1), so the strength of the Sun's heating faculty at perigee is very nearly $27\frac{27}{36}$ or $27\frac{3}{4}$. At apogee, the decrement to the heating faculty is one eighth of the perigeal increment of $\frac{3}{4}$, which comes to $\frac{3}{32}$, so the heating faculty is $26\frac{29}{32}$.

Offusius works through these calculations for several other examples, then shows mercy for his reader and concludes the chapter with an expanded version of his first table (our table 3.1), giving this time the amounts of each quality effused by each planet at five distances: the apsides, the mean distance, and two intermediate distances. We shall show greater mercy still and omit the table, which is in any case riddled with errors. The errors are unfortunate, since the remainder of the book consisted of refinements to this table at the end of chapter 6, based on other circumstances.

Effects of Other Circumstances

The altitude of a planet above the horizon affects the force of its effusions, as does its "delay" (*mora*), the length of time it is visible above the horizon. The quality of heat increases as a cubic function of altitude, and the other qualities as a quadratic function. Offusius prints not one but two tables of adjustments, one at intervals of $3\frac{3}{4}°$ of altitude, the other—more or less redundantly—at intervals of $5°$.

The adjustment for altitude is naturally proportional to the amount being adjusted, multiplying it by as much as a factor of 64 (for heat emanating from the zenith) or 16 (for other qualities from the zenith).[17] These adjustments are not to be neglected! They imply, incidentally, that the great table at the end of his chapter 6 gave the amounts of the four qualities effused by the seven planets at five distances *when they were on the horizon*.

The length of time a planet is visible above the horizon—its "delay" above the horizon—affects the strength of its influence in similar fashion. Offusius even uses the same tables for these two adjustments. The tables can be entered via a column labeled "Altitude" for the first

---

[17] The construction of the tables followed the rule that the multiplier equaled $(\frac{degrees}{30}+1)^3$ for heat, and $(\frac{degrees}{30}+1)^2$ for the other qualities.

adjustment, or via a column labeled "Delay" for the second. (This double use explains why the altitudes in one version of the previous table are given at intervals of $3\frac{3}{4}°$: that division results in twenty-four lines corresponding to delays of from one to twenty-four hours.) Hence a planet's delay above the horizon can multiply its effusion of heat by a factor of as much as 64 and its effusion of the other qualities by a factor of as much as 16.

Offusius does not address the variation of altitude that inevitably occurs during the planet's delay above the horizon. In one of his examples, the Moon, whose unmodified moistening faculty is 100, attains a strength of 121 at perigee; when it reaches an altitude of about 19° this strength is multiplied by $2\frac{41}{64}$ and becomes $322\frac{2}{3}$; and if it delays above the horizon for eight hours another factor of exactly 4 raises the strength of the lunar moistening faculty to almost 1,291. Of course, the Moon is not 19° above the horizon for all eight hours; the problem would seem to require an integration of the varying effect of elevation over the eight-hour period. It would be asking too much of a sixteenth-century astrologer's mathematical sophistication to expect his calculations to account for this. At any rate, we can infer that Offusius's original table had given the amounts of the four qualities effused by the seven planets at five distances when they were *momentarily* on the horizon.

A final consideration is the declinations of the different planets. If two planets have the same declination, or terrestrial latitude, it is as if they follow each other's footsteps across the sky from rising to setting. By doing so, they mutually strengthen one another. If the Sun is in 18° Sagittarius, and some other planet in 12° Capricorn without latitude— hence in the ecliptic—each has a declination of about 23° south. They follow each other's footsteps through the course of a day, each traveling along an arc parallel to the equator and 23° south of it.[18] Before explaining quantitatively how this intriguing circumstance affects planetary effusions, Offusius considers several other matters. One of these is a relation among the apparent sizes of the planetary bodies.

SIZES OF THE PLANETS

Offusius claims to have observed mathematical harmony in the apparent sizes of the planets. Starting with the luminaries, the Sun and Moon, he says he has often observed both of them very carefully,

[18] Offusius, *De divina astrorum facultate*, fol. 11$^{\rm v}$.

especially when they were at about their mean distances from us, and has found them to be the same apparent size. Their diameters, in fact, are both 33 minutes of arc—a measurement to which, he hastens to point out, the results of the ancients agree pretty well. His observations of the other planets are even more precise. "I have often observed Venus with all diligence, enthusiasm, and care (but when it was at the mean distances), and have found it to show us a diameter of $6\frac{2}{5}$ minutes or more." Jupiter he has observed "with no less keenness" and has found it to show a diameter a bit larger than $4\frac{2}{3}$ minutes; Mars, slightly more than 4 minutes; Saturn, $3\frac{1}{4}$ minutes; and Mercury, 3 minutes. The precise apparent diameters he claims for these planets are respectively one-fifth, one-seventh, one-eighth, one-tenth, and one-eleventh of the apparent diameters of the luminaries.[19] The apparent faces or "convexities," as he calls them, are consequently in the squared ratios to these diameters, as shown in table 3.5.

Not only are the numbers in column 3 perfect squares but the sum of these squares is exactly 360. If 1 more were added, he remarks, "for the Solar convexity which illuminates us," the total would be 361, which is another square number.

The discovery of these wonders releases an outpouring of laboriously circular calculation. Offusius started by noting that the diameter of Venus was a fifth, and its face a twenty-fifth, that of the Sun. Hence he computes the twenty-fifth part of the visible size of the face of the Sun, declaring that the quotient is in fact the visible size of Venus. From it he calculates the apparent diameter of the planet as $6\frac{7}{13}$ minutes; not surprisingly, the result differs little from the "observed" diameter with which he had started. Had he done this little exercise more accurately, it could only have proved his ability to convert consistently between disks and diameters.[20]

ASPECTS

Offusius goes on to explain why it is necessary to know the apparent sizes of different planets. When two planets are in conjunction or

[19] Ibid., fols. 15$^\text{v}$–16$^\text{r}$. These "observed" sizes are, of course, far too large and absurdly precise for naked-eye observations; Offusius obviously calculated them from his theory. Incidentally, Venus should be $6\frac{3}{5}$ minutes, not $6\frac{2}{5}$.

[20] Offusius used $\sqrt{10}$ as the approximate value of π. He was aware of the (much better) approximation $\frac{22}{7}$, but regarded it as *crassiusculam*, cruder: see fol. 16$^\text{r-v}$.

TABLE 3.5.
Convexities "Observed" by Offusius

1 Planet to Sun	2 as	3 to
Moon	1	1
Venus	1	25
Jupiter	1	49
Mars	1	64
Saturn	1	100
Mercury	1	121

opposition, their rays coincide and gain strength, each from the other. This collinear mixing of rays strengthens both, in proportions that can be calculated from the sizes of the apparent hemispheres, or convexities, of the sources. Hence these two aspects, conjunction and opposition, are of great importance in calculating the effects of planetary effusions. Offusius sees no physical reason, on the other hand, that sextile, quartile, or trine aspects might affect the rays emitted by planets, although he admits that these aspects are universally used by astrologers.

A similar but lesser mutual strengthening occurs when two planets have the same declination or terrestrial latitude, because then they "follow each other's footsteps" in their daily rotation around the Earth, as discussed above. One presumes that atmospheric disturbances—an important avenue for astrological influence—are particularly susceptible to this kind of influence because the rays of one planet follow the same path that those of the other followed a few hours before. Planets with the same declination thus benefit from a kind of delayed collinearity of rays.[21]

The planetary distances calculated earlier are essential in determining intervals of planetary influence. To calculate the interval of time during which a conjunction or opposition is efficacious, one needs to know the moments at which the rays become collinear within the

[21] Ibid., fols. 12ᵛ–14ᵛ; 18ʳ–19ᵛ; 21ᵛ–22ʳ.

region of interest. Disturbances in the weather, for instance, originate in the region of air. The flexibility (to choose a neutral word) of Offusius's theory of planetary distances is handy here, for it enables him to estimate the boundaries of the sublunary spheres of fire and air. The mean distance of each planet's sphere, by his theory, is $\frac{3}{8}$ that of the sphere inside it; why should not the same ratio hold between the spheres below the Moon? The mean distance of the Moon's sphere is $30\frac{3}{8}$ terrestrial diameters, so Offusius confidently puts the mean distance of the sphere of fire at $\frac{3}{8}$ of this, or about $11\frac{1}{2}$ terrestrial diameters; and the mean distance of the sphere of air at $\frac{3}{8}$ of that, or about $4\frac{1}{4}$ terrestrial diameters.

The outer boundary of ordinary, noncelestial matter is thus somewhere between $11\frac{1}{2}$ terrestrial diameters and the inner boundary of the lunar sphere at $25\frac{7}{8}$ diameters—probably, he thinks, around $18\frac{1}{2}$ terrestrial diameters. (Offusius is not altogether certain how to characterize the region, if there is one, between the region of fire and the innermost sphere of the Moon. It is most likely filled, he thinks, with matter that is corruptible yet that participates somewhat in the virtue of the heavens. Perhaps comets originate in this region.)

Knowing the dimensions of the sublunary spheres enables Offusius to determine the moment at which the line joining two planets, the line along which their effusions of virtue are combined and mutually strengthened, begins to impinge upon the various regions of the sublunary world. Figure 3.1 shows the critical geometry, simplified slightly from the original figure.

At the center of the figure is a spherical region composed of the elements earth and water—the planet Earth, in fact. Around it lie the regions of air and of fire, surrounded by the corruptible yet heavenlike sublunary region whence come comets, and finally the inner surface of the sphere of the Moon. The outermost circle represents the path BHK of the Moon as it circles the sublunary world. At the top of the figure the innermost planet, Venus in the cosmology of Offusius, is at A.

When the Moon is at B its rays join with those of Venus along ABC, and the effusions of both planets are accordingly strengthened at all points along that ray—which just barely brushes the edge of the sublunary world at C. At a later time, when the Moon has moved to D, the ray of mutually strengthened influence ADE passes through a fairly large portion of the sublunary world and just begins to graze the sphere of fire. One presumes that if a comet is forming somewhere in

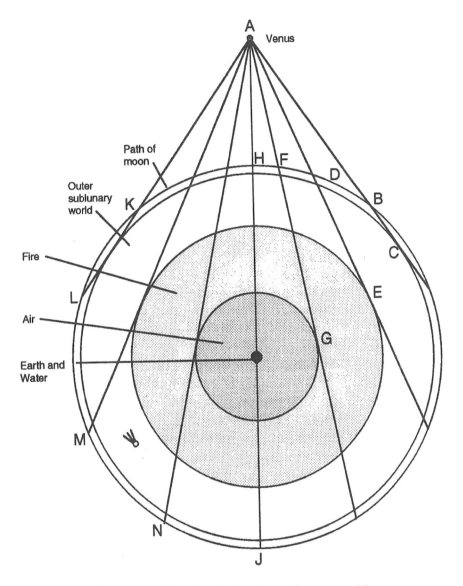

FIGURE 3.1. Rays impinging the sublunary world

the outer sublunary world at the moment when the ray sweeps through, it will be a much moister comet than most, from the combined effects of the lunar and Venusian rays.

When the Moon reaches F, and the ray joining it and Venus impinges on the sphere of air at G, the effects of this ray begin to appear in the weather, growing in strength as the ray sweeps into the middle of the airy region, and continuing as long as it remains therein. The effects of the conjunction diminish, or rather withdraw from the Earth, as the Moon continues on its path. Soon it reaches K, where the ray joining it to Venus leaves the sublunary world altogether. But soon it reenters; and as the Moon is carried around to L, M, and N the combined and strengthened influences of the two planets' opposition reach the outer sublunary world and the spheres of fire and of air. Clearly the Moon, by virtue of its proximity to Earth, strengthens the effusions of every other planet very often, in one region or another of the sublunary world, by conjunction or opposition. Consequently the position of the Moon is of the greatest importance in calculating the influence of other celestial bodies.

OFFUSIUS AND KEPLER

As nearly as one can tell from this single work (one of several that he claims to have written, but the only one I have been able to locate), Offusius's purpose in harmonic speculation is primarily astrological. If he is also motivated by admiration for the Creator's works, or by belief that the fundamental principles of geometry can be discerned in all phenomena partaking of Reason, he feels no need to introduce those motivations. He wants simple mathematical relations to use as tools in elaborating his theories of planetary influence. The supposition of beautiful and elegant proportions among heavenly dimensions provides him with an inexhaustible supply of such relations.

Elegant or harmonic relations served a rhetorical purpose as well. Offusius, like most writers on the subject (including Kepler), takes it for granted that readers are inclined to believe in the reality of proportions that are intrinsically beautiful. The enduring attraction of the "harmony of the spheres" is no doubt due in large part to a common wish that the world should be easy to understand and constructed explicitly for our admiration and benefit.

But there was a more specific benefit in the methods of Offusius. His characteristic way of producing relationships at pleasure when he needed them, and working out in detail calculations based on those relationships, was particularly suited to the astrologer's trade. He had the habit, common in his age and indeed all ages, of cautiously drawing a conclusion and then recklessly building upon it. He took questions about the proportions of the heavens as deserving of serious investigation, but he had no idea how to conduct such an investigation.

Putting aside questions of ability and method, one can still detect an important difference between the beliefs of Offusius and those of Kepler regarding harmonic principles in the world. Offusius did not believe in the physical efficacy of harmonic proportions. He did not think aspects were important, aside from conjunction and opposition, and he believed in those aspects only because the rays from planets in conjunction or opposition coincided and thus strengthened each other, by brute force as it were. This was quite different from Kepler's theory that human or planetary souls were excited by perceiving rays that joined at a harmonic angle such as 60° or 90°. Offusius thought that the world had been created according to harmonic principles, but he did not believe that an eternal harmony permeated the day-to-day workings of nature, nor did he believe that our souls are stimulated by the perception of harmony as such.

Distances to the Planets

OF ALL THE MANIFESTATIONS of harmony in the physical world, the most sublime were surely those hiding in the proportions of the heavens. It took no great originality to suspect that the heavens had been laid out according to some plan, and no great curiosity to want to discover that plan. The problem was only to discern the proportions themselves.

Which proportions were of interest depended on the way in which the heavens were laid out. So long as one viewed the heavens geocentrically, planetary distances from Earth were the principal dimensions of the large-scale plan. If one considered the Sun as the center of the system, distances from the Sun became significant. Calculating the proportions among these distances was possible in either system, given suitable assumptions.

IN GEOCENTRIC ASTRONOMY

For harmonic theories of the heavens, relative distances to the planets were needed. Harmonies depended only upon proportions among magnitudes, so ignorance of the absolute distances was not a problem. Harmonic speculation did require a knowledge, or at least an opinion, about the proportions among the distances. And opinion it was; prior to Copernicus, calculation of the relative distances to the planets was one of the most delicate parts of mathematical astronomy, and one of the least satisfactory. The relative sophistication of the Ptolemaic planetary models did not extend to comparisons among models for different planets, because the models were independent of one another. The calculations were made nonetheless, by one stratagem or another.

In the *Almagest* itself, Ptolemy compared the distances only of the Sun and the Moon (and he erred there by a factor of 20, because the calculations were extremely sensitive to small errors in observed angles). Yet Arabic and Latin writers of the centuries between Ptolemy and Copernicus did not hesitate to demonstrate the proportions among the greatest, mean, and least distances to all the planets. Their estimates, moreover, showed great consistency of method and even of result. Discrepancies among them arose from particular, identifiable refinements in a universally accepted method.

The pedigree of this method, once something of a mystery, could not have been better, for Ptolemy himself explained the method in a late work known as the *Planetary Hypotheses*. The crucial sections, containing derivations of the relative distances to the planets, were lost in all surviving Greek manuscripts of the *Planetary Hypotheses* but survived in Arabic and Hebrew versions, from which B. Goldstein published them in 1967.[1] Medieval writers who confidently described the proportions of the planetary spheres followed the authentic tradition of Ptolemy.

The method combines models of mathematical astronomy with crude physical theories. The physics is minimal, the simplest possible translation of technical planetary models into three-dimensional solid geometry. It is based on the very plausible assumption that heavenly motions arise from the rotations of rigid spheres. Motions in the heavens are obviously periodic, and rotating spheres are the most obvious way of generating periodic near-circular motions in three dimensions. The entire heavens are assumed to be enclosed in a sphere called the *primum mobile*. This carries the "fixed" stars and the "wandering stars," which are the planets, the Sun, and the Moon, with it in its daily rotation from east to west around the Earth, which rests immobile at the center. Each wandering star is individually carried back from west to east, much more slowly, by its own system of spheres. The Ptolemaic mechanisms, involving spheres rotating within spheres, account for the recurring periods of retrograde motion, from east to west, that are observed in all the wandering stars except the Sun and Moon.

The arrangement of rigid spheres in a Ptolemaic planetary model is shown in figure 4.1 (not to scale for any particular planet).[2] The model is a three-dimensional nesting of spheres; hence each circle in the figure must be understood as the surface of a sphere. At the center of the model—and the universe—is the Earth, *T*. The outer surface of

---

[1] Goldstein, "The Arabic Version of Ptolemy's Planetary Hypotheses."

[2] A concise history of these models is in van Helden, *Measuring the Universe* (Chicago, 1985), chaps. 3 and 4.

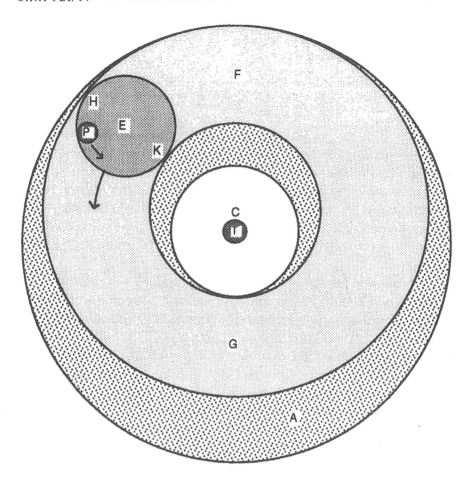

FIGURE 4.1. Solid-sphere model

sphere *A* is centered on *T*. Off-center within it is another sphere *FG*. This is the *eccentric sphere* or simply the eccentric, and point *C* is the center of the eccentric. Embedded within the eccentric sphere is a smaller sphere, the *epicycle*, centered on *E*, which is carried eastward (counterclockwise) by the rotation of the eccentric, *FG*. When the epicycle is most distant from the Earth, at *F*, it is said to be at the apogee of the eccentric; when closest to Earth, at *G*, it is said to be at perigee of the eccentric. Meanwhile the epicycle also rotates counterclockwise around its own center, *E*, carrying the planet *P* embedded within it. When it turns so that the planet is farthest from Earth, at *H*, the planet is said to be at apogee of the epicycle, and when closest to the Earth, at *K*, it is at perigee of the epicycle. The planet's greatest distance from Earth is obtained when its epicycle is at apogee of the eccentric, and it is itself simultaneously at apogee of the epicycle. Its closest approach to Earth occurs when it and its epicycle are both at their respective perigees.

A solid-sphere model such as this accounts for the components of Ptolemy's mathematical planetary model because it *is* that model, fleshed out into three dimensions. Figure 4.2 shows the mathematical model, illustrating the subtle differences between the two.[3] A mathematical diagram drawn by an astronomer represents the eccentric sphere by a circle midway between its inner and outer surfaces, circle *FEG*, centered on a point *C* that is eccentric, or removed from the center of the universe, by a distance *TC* = *e*, called the eccentricity. It likewise represents the epicycle by a circle passing through the center of the planet *P*. This representation shows what is important for the astronomical problems of calculating angles and motions. The spherical model of figure 4.1 dresses up the astronomical model to deal with the physical problem of accounting for the motions.

The relative dimensions of the parameters *TC*, *CE*, and *EP* in figure 4.2 can be calculated from astronomical observations. These suffice to determine the proportion between the inner and outer surfaces of the assembly in figure 4.1, if one neglects the radius of the planet *P* itself, which is certainly a very small quantity. The inner surface is at the distance of the planet's closest approach, when the epicycle is at perigee of the eccentric and the planet is at perigee of the epicycle. This least distance equals *CE* − *TC* − *EP*, the radius of the eccentric minus the eccentricity minus the radius of the epicycle. The distance of the outer surface from the center is similarly *CE* + *TC* + *EP*, the sum of the two radii and the eccentricity.

Both the outer and inner surfaces of the nest of spheres in figure 4.1 are centered on *T*, which is also the center of the universe. Hence it is easy to build a nest from the models for all the planets, fitting each into the central cavity of that superior to it. The proportion between these two surfaces is different for each planet but can be calculated astronomically for each as $\dfrac{CE - TC - EP}{CE + TC + EP}$. By assuming that no unnecessary empty space exists between the spheres, Ptolemy nested the planetary models snugly and calculated all of the distances between them and the Sun.

---

[3] For a detailed analysis of this seldom-discussed material see Swerdlow, "Pseudodoxia Copernicana."

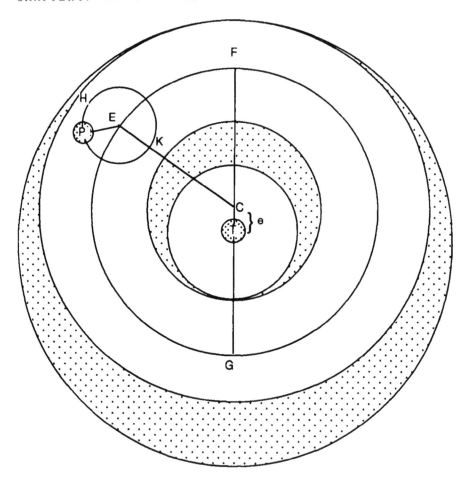

FIGURE 4.2. Geometry of a solid-sphere model

This last assumption—that no space is wasted—is perhaps philosophical rather than physical or astronomical, but it is compelling for accidental, although seemingly scientific, reasons as well. Another argument in the *Almagest*, entirely independent of this and based on the size of the Earth's shadow during lunar eclipses, yields the distances to the Moon and the Sun in terms of terrestrial radii.[4] The computed solar distance is 1,210 terrestrial radii (t.r.). In the *Planetary*

---

[4] See Neugebauer, *A History of Ancient Mathematical Astronomy* (Berlin and New York, 1975), 1.103–11. Ptolemy's estimated lunar distance was surprisingly accurate, due in part to "the accidental interplay of a great number of different inaccuracies of empirical data and of computations" (p. 106). His solar distance, on the other hand, was too small by a factor of 20.

Hypotheses, Ptolemy calculates (by nesting the models for the Moon, Mercury, and Venus around the moon's distance) that the outermost sphere of Venus is at a distance of 1,079 t.r. Now, if one assumes that 1,210 t.r. is the *mean* distance to the sun—a detail left unspecified in the *Almagest*—then the distance to the innermost sphere of its model would be 1,160 t.r., remarkably close to the value calculated from the nesting of spherical models. A trifling adjustment to the observational data would lead to identical distances from the two lines of argument.

No doubt this remarkable, if entirely fortuitous, confirmation of the distances implied by the solid-sphere models contributed a great deal to their acceptance. At any rate, the Ptolemaic distance estimates were widely disseminated in the Arabic treatises through which medieval Europe learned of ancient science. The assumptions on which they rested—in particular the assumption that no space was wasted in the heavens—were both natural and consistent with much larger patterns of thought. The distances themselves were not of the sort that could be easily measured, but Ptolemy's independent calculation of the solar distance provided confirmation. The general scheme of calculating distances to the planets by nesting spheres of the requisite sizes was soon well established. Numerous modifications of it were proposed; some writers even embellished it by leaving room for the estimated diameters of the planets themselves. There was surprisingly little disagreement on the *apparent* diameters of the planets, each writer apparently feeling inadequate to correct the naked-eye estimates of his predecessors. (In fact, these estimates were far too large; the planets are essentially point sources of light smeared by the Earth's atmosphere.)[5]

As representative of the distances implied by solid-sphere models, we display those of al-Battānī in table 4.1.[6]

The salient features of this and related geocentric schemes are that the greatest distance of each planet's sphere equals the least distance of the next higher planet's sphere, and that the sphere of the fixed stars is assumed to lie just beyond that of Saturn. No space has been wasted by the creator of this universe. It is worth remarking that the thickest spheres (proportionally) are those of Venus and Mars. This is because the epicycle of a planet's synodic anomaly largely determines the thickness of its sphere, as shown in figure 4.1, and that epicycle is in

[5] Van Helden, *Measuring the Universe*, chap. 4.
[6] Ibid., p. 32.

69

TABLE 4.1.
Planetary Distances of Al-Battānī

1	2	3	4
	Absolute distances in terrestrial radii:		
Body	Least	Mean	Greatest
Moon	$33\frac{1}{2}+\frac{1}{20}$	$48\frac{5}{6}$	$64\frac{1}{6}$
Mercury	$64\frac{1}{6}$	$115\frac{1}{2}$	166
Venus	166	618	1,070
Sun	1,070	1,108	1,146
Mars	1,146	4,584	8,022
Jupiter	8,022	10,473	12,924
Saturn	12,924	15,509	18,094
Fixed stars		19,000	

the proportion of the Earth's heliocentric orbit to that of the planet. The Earth's annual revolutions about the Sun had their greatest effect on the apparent motions of the planets whose orbits are nearest in size to that of Earth, namely Venus and Mars.

As we shall see in chapter 7, when Kepler reconstructed the final chapters of Ptolemy's *Harmonics* he assumed that a scheme something like this underlay Ptolemy's harmonic cosmology.

In Heliocentric Astronomy

The publication of Nicholas Copernicus's book *De revolutionibus orbium coelestium* in 1543 initiated a radical transformation of astronomy, although this did not become apparent for several decades. For the first time the proportions of the planetary system appeared directly in the models of mathematical astronomy. The dimensions of the Ptolemaic heavens, and hence the proportions among those dimensions, had rested on an accidental agreement between one method based on assumption and another based on particularly delicate measurements. Copernicus's heliocentric rearrangement of the Ptolemaic

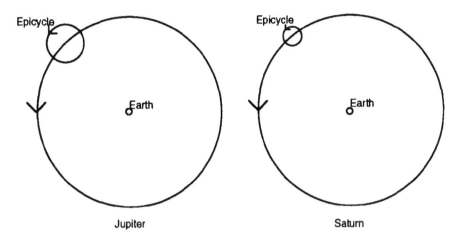

FIGURE 4.3. Geocentric second-anomaly models for Jupiter and Saturn

models revealed what had been hiding in the old models all along: the relative sizes of the planetary spheres.

De revolutionibus itself consisted almost entirely of mathematical astronomy, an old and highly technical science. Its pages were filled with observational reports, greatly variable in quality, which had been passed down through the centuries; and with diagrams, demonstrations, and tables of numbers. Such contents made for dry reading, but as an attempt to rederive all the parameters of mathematical astronomy the book found an audience among contemporary astronomers. One attentive reader was Michael Maestlin, who taught astronomy to the young Kepler at Tubingen. The most impressive part of the new hypotheses, as Maestlin realized, was Copernicus's unification of the models for the second (or "synodic") inequalities of all the planets into the single model of a moving Earth. On the other hand, the most disturbing implication, to an astronomer, was an enormous expansion in the sphere of the fixed stars.

Let us see what this means. Figure 4.3 shows pre-Copernican models for the synodic inequalities of the planets Jupiter and Saturn, omitting the much smaller apparatus for the first (zodiacal) inequalities. The epicycle is drawn approximately to scale in each model: Jupiter's is about one-fifth the radius of the main circle, and Saturn's about one-tenth. These ratios are calculated from inequalities observed in the motions of the two planets. The two models, however, are entirely distinct. Physical assumptions about closely nested spheres,

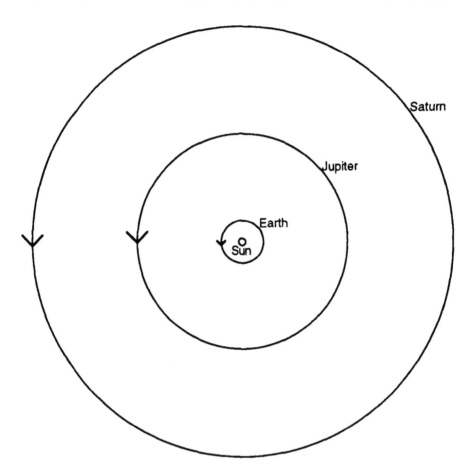

FIGURE 4.4. Heliocentric second-anomaly model for Jupiter and Saturn

such as those discussed above, were not a part of mathematical astronomy before Copernicus. There was no mathematical basis for
comparing the sizes of the two models in figure 4.3.

The corresponding Copernican model for the synodic inequalities
of the same two planets is in figure 4.4. It is a single model and would
be a single model still if it included all six planets. The anomalies in
motion previously attributed to the epicycles are here accounted for
by the single motion of the Earth around the Sun. As before, the ratios
among the circles are calculated from the observed motions. Indeed
they are the same ratios as before—but in this model, they are ratios
among the circles of the different planets. Earth's orbit is about one-
fifth the radius of Jupiter's, and about one-tenth the radius of Saturn's.

TABLE 4.2.
Copernicus's Planetary Distances

1	2	3	4
	Relative distance in astronomical units:		
Body	*Least*	*Mean*	*Greatest*
Mercury	0.2627	0.3763	0.4519
Venus	0.7018	0.7193	0.7368
Earth	1.0	1.0	1.0
Mars	1.3739	1.5198	1.6657
Jupiter	4.9802	5.2192	5.4582
Saturn	8.6522	9.1743	9.6963
Fixed stars		Immense	

A single assumption, that the Earth moves around the Sun, thus permitted followers of Copernicus to calculate the relative distances among all the planets.

Table 4.2[7] gives planetary distances from the Sun as derived from the heliocentric models of Copernicus. The natural unit of measurement in these models is the radius of the Earth's orbit. Distances are measured from the center of that orbit, since Copernicus constructed the models for the other planets around that center.

The change in models also implied, unfortunately, that the fixed stars (not shown in either figure) had to be vastly more distant from Earth than had ever been imagined. The stars were called "fixed" precisely because they were not seen to move; but if the motion of an observer on Earth around the Sun made no perceptible difference in the direction to a star, the star must be many, many times more distant than the Sun. Tycho Brahe estimated that his massive instruments would have detected any apparent motion as large as 1' of arc in the

---

[7] Adapted from ibid., p. 44, who uses for Mercury's distances those of Maestlin as given by A. Grafton, "Michael Maestlin's Account," p. 549. I have changed the extreme distances for Earth to 1.0 in order to use the center of the Earth's orbit consistently as the center for all the planets.

fixed stars. Since he had detected no such motion, he calculated that Copernicus's hypothesis required the stars to be distant by nearly eight million terrestrial radii.[8] Such extravagance was hard to accept in the face of a scientific tradition that put the stars at a distance of around twenty thousand terrestrial radii.

Equally offensive were the implications of such distances for the size of the fixed stars. Ptolemy had estimated that a star of the first magnitude had an apparent diameter one-twentieth that of the Sun: about a minute and a half of arc. His successors had generally accepted that estimate; certainly no one had proposed that the angular sizes of all the fixed stars were essentially zero. Today we know that the "apparent" diameters were due to the scattering of starlight in Earth's atmosphere; but if one took such estimates at all seriously, the Copernican distances to the fixed stars translated into gigantic estimates of the sizes of those stars. A star whose apparent diameter was a minute and a half at a distance of eight million terrestrial radii would have a diameter of nearly thirty-five hundred terrestrial radii and consequently a volume over five billion times that of the Earth. Such numbers—for a single star in the heavens—were simply beyond belief.

Tycho was, of course, correct in inferring that the Copernican hypotheses required an enormous distance to the fixed stars; the nearest star is indeed well beyond eight million terrestrial radii. He can hardly be blamed for shrinking from the further Copernican implication that almost all the volume within the sphere of the fixed stars was wasted outside of Saturn's orb, serving no purpose. This would have been absurd, in direct contradiction to the governing assumption of all previous theories of heavenly distances, and contrary to a good part of natural philosophy.

[8] Van Helden, *Measuring the Universe*, p. 51.

The Polyhedral Theory of the
Mysterium cosmographicum

KEPLER PUBLISHED his first attempt to explain the dimensions of the solar system in 1596, when he was twenty-five years old, in a book of fewer than a hundred pages entitled *Introduction to the cosmographic treatises, containing the cosmographic mystery concerning the remarkable proportion of the heavenly spheres, and concerning the genuine and proper causes of the number, magnitude, and periodic motions of the spheres, demonstrated by means of the five regular geometric solids.* Today it is generally known as the *Mysterium cosmographicum.*[1] Most of the larger problems that concerned Kepler throughout his career as an astronomer were raised in this book—raised, indeed, in its title. Our particular concern in this chapter is with the polyhedral theory, which explained the answers to two of these problems: the number of planets and the sizes of their spheres.

The polyhedral theory explained the number and sizes of the six planetary "spheres" by nesting the five regular geometrical solid figures, or polyhedra, between them. Its use of the idea of planetary spheres is a puzzle, since Kepler had already rejected the traditional models of astronomy, involving actual spheres carrying the planets through the heavens. In chapter 16 of the *Mysterium* he called the idea of rigid planetary spheres "absurd and monstrous."[2] Yet though he denied the reality of rigid, physical spheres, he conceived of the universe, like every astronomer in the 1590s, in terms of spheres within spheres. This world-picture had lost for Kepler, by 1605, every connection with what he thought was taking place in the heavens, but the language and presumably the image of planetary spheres remained with him to the end.

[1] *Prodromus dissertationum cosmographicarum* (Tubingen, 1596). Maestlin saw the book through the press, whence it finally emerged in March 1597 (*G.W.* 13.108). The modern edition is in vol. 1 of *G.W.* There is an English translation by A. M. Duncan, with notes by E. J. Aiton.

[2] *G.W.* 1.56.17. See also J. V. Field, "Kepler's Rejection of Solid Celestial Spheres," pp. 207–11.

Origin of the Polyhedral Theory

Kepler recounted the origin of his polyhedral theory in his preface. On 19 January 1595 he had been lecturing on the subject of Great Conjunctions, a matter of much astrological importance. A Great Conjunction occurs when Jupiter and Saturn pass one another in the zodiac. These are the two slowest planets, so their conjunctions are rarer than those of any other pair of planets. On average, Jupiter moves 30° in a year, and Saturn 12°. Hence Jupiter gains about 18° per year and, on average, overtakes Saturn every twenty years. In the twenty years since the previous conjunction, Saturn has moved about two-thirds of the way around the zodiac, and Jupiter has made a complete revolution and two-thirds of another. The locations of these mean Great Conjunctions therefore jump forward by two-thirds of a circle (more precisely 243°) every twenty years.

Three successive mean Great Conjunctions mark very nearly an equilateral triangle in the zodiac, since the third occurs only 9° beyond the first. If one continues to draw these near–equilateral triangles, one eventually obtains a figure resembling figure 5.1, which Kepler drew for his class and later reprinted in the preface to the *Mysterium cosmographicum*.

It is impossible not to notice a smaller circle, implicit in figure 5.1, to which all the lines are tangent. The radius of this circle inscribed in the triangles is very nearly half that of the outer circle, because the perpendicular distance from the center to the side of an equilateral triangle is half the distance from the center to a vertex. But as Kepler knew, the radius of Jupiter's path is about half that of Saturn's path. A diagram of the Great Conjunctions of Jupiter and Saturn showed also the proportion between the spheres of Jupiter and Saturn. Moreover, the equilateral triangle was the first regular polygon, as Jupiter and Saturn were the first planets.

"Immediately," Kepler continued—one wonders if he was still standing in front of his class—he went on to compare Mars and Jupiter's ratio of solar distances to the square, intending to compare that of Earth and Mars to the pentagon, that of Venus and Earth to the hexagon, and so on. His hopes were quickly dashed, however, since the ratio between Mars and Jupiter is far greater than the ratio between the circles inscribed and circumscribed to a square. He tried combining the square with the triangle and the pentagon, but with little success

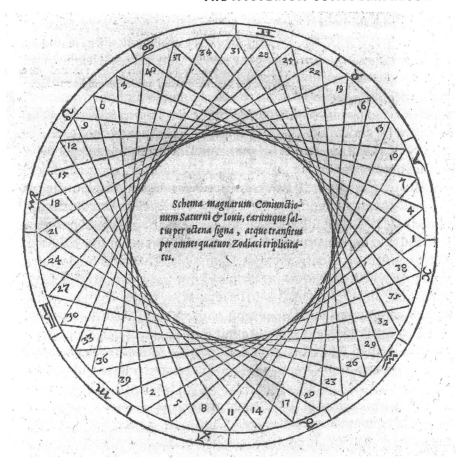

FIGURE 5.1. The progression of Great Conjunctions

and less hope. There would be no end to such attempts, as he wrote: using the polygons in order, as he was, he would never reach the Sun, for geometry offered an infinite number of regular polygons.

Only six planets revolved around the Sun. As his initial excitement subsided, Kepler considered this number carefully. By comparing the circles inscribed and circumscribed to polygons, one could derive ratios among twenty, or a hundred, planets; but only six were needed. Any theory that accounted for the relative sizes of the planetary orbits would have to produce the five ratios Jupiter:Saturn, Mars:Jupiter, Earth:Mars, Venus:Earth, and Mercury:Venus—and no more. Yet the figures pleased him as an explanation of the spheres' distances, because they were geometrical, a matter of pure quantity and as such "prior to the heavens." Quantity had been created on the First Day, according to the biblical account in Genesis, while the heavens were not established until the Second Day.[3] If only, he thought, he could single out five figures among the infinite number of possible ones, and use those five figures to explain the proportions among the six planetary spheres. Concentrating on this wish, Kepler had an inspiration: what were polygons, plane figures, doing among the spheres of the heavens? He should use the polyhedra, solid figures, instead.

With this realization the problem solved itself. Kepler had been considering the infinity of regular polygons; but every geometer knew that there were precisely five regular polyhedra. Regular polyhedra were solid figures composed of a single type of regular polygon, having all their faces and all the solid angles between their faces identical.

Euclid had demonstrated in the final proposition of the *Elements* that only five such polyhedra were possible. The solid angles of a regular polyhedron had to be formed from the juncture of at least three faces, and these faces had to be regular polygons. The equilateral triangle was the first regular polygon, with an angle of 60° at its vertex. Three such triangles made a solid angle from the regular *tetrahedron*, a pyramid-shaped figure composed of four equilateral triangles in all. Four equilateral triangles made a solid angle from the regular *octahedron*, which had a total of eight faces. Five made a solid angle from the regular *icosahedron*, which had twenty equilateral triangles as faces. Six equilateral triangles could not make a solid angle: six 60° angles made 360° in all, so the triangles would spread flat, forming a plane figure (a hexagon) rather than a solid angle. Turning to the square with its vertex angle of 90°, one obtained the solid angle of a *cube* from three squares; but four or more squares in a solid angle were impossible. The next candidate for the polyhedral face was the regular pentagon, whose vertex angle was 108°. Three of these made up an angle of the regular *dodecahedron*, formed from twelve pentagons. Four or more pentagons could not make a solid angle; and even three regular hexagons were impossible. Regular polygons with seven or more sides were a fortiori impossible in a solid angle, so the five figures enumerated above were the only possible regular polyhedra.

The regular tetrahedron, cube, octahedron, dodecahedron, and icosahedron were therefore precisely what Kepler wanted: five solid figures singled out, among all possible solid figures, by their mathematical beauty. Although composed of planes, they resembled the sphere,

[3] *G.W.* 1.12.9–13.1.

most perfect of shapes. All their faces were equidistant from the center (and hence were tangent to a single inscribed sphere). All their angles were equidistant from the center (and defined a sphere that circumscribed the figure). The ratio between the radii of these two spheres was thus well defined, and characteristic, for each of the figures. A sphere's perfection could not be constructed from planes, but it was most nearly approached in these five figures.

The tetrahedron had the greatest ratio of spheres (the ratio farthest from equality), and it was exactly 1:3. The cube and octahedron had equal ratios of $1:\sqrt{3}$. The dodecahedron and icosahedron likewise had equal ratios, the closest to equality of any of the regular figures, and approximately equal to 4:5.

ARRANGEMENT OF THE FIGURES

After explaining, in the preface, how he had realized that the regular polyhedra were apt to solving the cosmographic mystery, Kepler mentioned that he had assigned each of them to the appropriate interval between planets before he had any knowledge of what the best arrangement would turn out to be. By this he did not mean what one might expect, that he had fit the figures into the gaps between planetary spheres without knowing the size of those gaps, and the figures had turned out to be just the right sizes. Quite the opposite.

Kepler instead remarked, with some pride, that he had assigned each polyhedron to its proper place among the spheres before he understood the order of dignity or priority (*praerogativae in ordine*) of the figures. On the basis of his knowledge merely of the planets' relative distances, he had correctly nested the five solid figures in their natural order, and he had never found it necessary to revise this order.[4] Later, in chapters 5–8 of the *Mysterium* and again in book 5 of the *Harmonice mundi*, Kepler presented his polyhedral theory in the opposite order, explaining first the natural order of the figures and only then showing that this order agreed with the proportions in the heavens. His ingenuous admission here that he arranged the figures to fit the distances disarms one's skepticism, converting a claim of order found in the heavens into a less striking one of order in the more subjective "natural" arrangement of the regular polyhedra.

---

[4] Ibid., 1.13.12–16.

The natural arrangement of the regular polyhedra was based, first of all, upon a division of them into two classes. The *primary* figures were the cube, the tetrahedron, and the dodecahedron. Their solid angles were formed by the juncture of three faces, and those faces differed among the three figures. The cube had square faces, the tetrahedron triangular faces, and the dodecahedron pentagonal faces. The *secondary* figures, the octahedron and the icosahedron, had angles formed by four or five faces, and all their faces were triangles.

Kepler had found other distinctions between the classes as he developed the polyhedral theory. The primaries had to be suspended through the center of a face and the opposing face (or angle) if they were to revolve "harmoniously" (*concinnè*), whereas the secondaries had to be suspended through opposing angles. This peculiar and vaguely stated mark of distinction referred to what is today called rotational symmetry. The primary figures have a higher order of rotational symmetry about an axis through the center of a face, while the secondary figures have a higher order of rotational symmetry about an axis through an angle. The tetrahedron answers to both descriptions, since it has an angle opposite to the center of each face. The primary figures, he continued, properly rested on a face, whereas to the secondary figures it was more appropriate to "hang," presumably from an angle. A primary figure standing on one of its angles, or a secondary figure resting on one of its sides, was repellent to the eye.[5]

The most fitting order of the figures (and the order that best fit the observed distances) was thus the following. The cube occupied the outermost interval, between Saturn and Jupiter. This was the interval closest to the fixed stars, which were the most dignified part of the world "aside from Earth" in a geometrical sense, as the circumference of a circle was its most dignified part aside from the center.[6] Kepler had placed it there because it fit in that interval, as we have seen; but the octahedron fit equally well, if a good fit had been all that was desired. When first placing the figures among the planetary spheres he had put the cube in this most dignified place, if we may believe him; and there is no reason to doubt it. In pondering the matter he had come to the conclusion that the cube was indeed foremost among the regular polyhedra, for a number of fairly obvious reasons. Its parts were all perpendicular or parallel to one another, and corresponded—indeed, represented—the three dimensions of space. It rightly occupied the first interval.

[5] See Field, *Kepler's Geometrical Cosmology* (Chicago, 1988), p. 55.

[6] *G.W.* 1.31.14–16. Kepler slips here into geocentric language.

The next interval inward, that between Mars and Jupiter, belonged to the tetrahedron. This had been obvious, since the tetrahedron's ratio of spheres, 1:3, was far larger than that of any other regular polyhedron, and the orbital radii of these two planets were similarly disparate. Upon reflection Kepler had decided that this was as it should be; among the regular figures the pyramid "almost dares to contend with the cube for primacy."[7] It was simple in form; the other perfect figures could be cut into many copies either of it or of its irregular variants; and it had various other attributes that confirmed his original judgment that it belonged between Mars and Jupiter.

The two intervals among the planets Mars, Earth, and Venus were approximately equal. On empirical grounds, both the dodecahedron and the icosahedron fit these intervals equally well, since the proportions between the radii of the inscribed sphere and the circumscribed sphere were identical for these two figures. There could be no question which figure to use, however. Two of the three primary figures, the cube and the tetrahedron, had already been assigned to the first two intervals, counting from the outside. The third primary figure was the dodecahedron, so Kepler confidently placed it between Mars and Earth, where it fit fairly well. The sphere of Earth, therefore, was at the division between primary and secondary figures.

The two secondary figures remained, the icosahedron and the octahedron. Many reasons indicated that the octahedron ranked first in this pair. The octahedron was carved out of the cube, the icosahedron out of the less-noble dodecahedron. (An octahedron can be constructed by connecting the midpoints of the faces of a cube, and an icosahedron by connecting the midpoints of the faces of a dodecahedron.) Indeed there were strong resemblances between cube and octahedron, and between dodecahedron and icosahedron. The octahedron was equally tall (*aequealtum*) as the cube, the icosahedron equally tall as, again, the less-noble dodecahedron.[8] Each figure had as many angles as its companion figure had faces, and as many faces as its companion figure had angles. Kepler displayed these

[7] Ibid., 1.32.15.

[8] These pairs of figures were "equally tall" when considered as supporting pillars of the celestial spheres. Each of the regular polyhedra could be thought to stand upon its angles within the circumscribed sphere and to support with its faces the inscribed sphere (or vice versa, but Kepler seems to have preferred the larger sphere as the base). Within this image, the cube and octahedron were equally tall because they supported the inscribed sphere at the same proportional distance from the circumscribed sphere.

TABLE 5.1.
Relations among the Regular Polyhedra

Figure	Face type	Faces	Sides	Angles	Inscribed sphere
Cube	Square	6	12	8	Intermediate
Octahedron	Triangle	8	12	6	Same as cube
Dodecahedron	Pentagon	12	30	20	Largest
Icosahedron	Triangle	20	30	12	Same as dodecahedron
Tetrahedron	Triangle	4	6	4	Smallest

relations in a table (table 5.1),[9] which makes clear the resemblances between the two figures in each pair.

The octahedron, then, was the nobler of the two secondary figures. Yet it did not follow the last of the primary figures in the next available interval, which was that between Earth and Venus. These seemed reasonable to Kepler. The two orders of figures differed, and so their heads looked in different directions, the cube outward and the octahedron inward. The cube's position nearest the fixed stars was the most dignified, but the next-best position was surely that nearest the Sun. The octahedron occupied the interval between Mercury and Venus, and the icosahedron that between Venus and Earth.

Figure 5.2 shows all the spheres and the polyhedra between them, in an ornate figure that Kepler published in the *Mysterium cosmographicum*.

DISTANCES IMPLIED BY THE POLYHEDRAL THEORY

The polyhedral ratios of the radii of inscribed to circumscribed spheres were as follows: for the tetrahedron, 1:3; for both the cube and the octahedron, $1:\sqrt{3}$, or about 577:1000; and for both the dodecahedron and the icosahedron, $1:\sqrt{15 - 6\sqrt{5}}$, or about 795:1000. Kepler compared these with the ratios between planetary spheres from book 5 of Copernicus's *De revolutionibus orbium coelestium*. A number of technical questions arose in this comparison.

[9] *G.W.* 1.28.34–39.

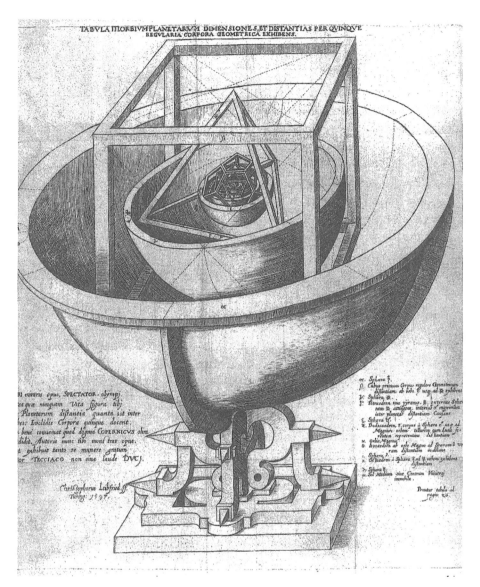

FIGURE 5.2. The polyhedral theory

Copernicus had built his models around the center of the *orbis magnus*, or great sphere, the sphere that carried the Earth. This center, which we call the mean sun, does not coincide with the Sun itself, because the Earth's orbit is eccentric. The orbits of the other planets are likewise eccentric, whether considered with respect to the mean sun or to the Sun itself; but the amounts of eccentricity vary depending upon which center is used. Since these eccentricities determine the "thickness" of the planetary spheres, they affect the nesting of the polyhedra among the spheres. Kepler, whose technical skills as an astronomer were still modest, sought the aid of his teacher, Michael Maestlin, who explained this and other matters to him in a series of detailed letters throughout 1596 and early 1597.[10]

Maestlin also raised the issue of whether the "sphere" of a planet needed to be large enough to include room for the epicyclic machinery needed to move the planet. This question arose from the physical models assumed to underlie astronomy. Figure 5.3 shows two positions of the epicycle that Copernicus had used to represent the motions of a superior planet. The epicycle E and the planet P are depicted at greatest and least distance from the mean sun S. When the planet is at aphelion of the eccentric, at the top of figure 5.3, it is always at perihelion of the epicycle, as shown; and when at perihelion of the eccentric it is always at aphelion of the epicycle. (Copernicus had arranged his models in this way to obtain the effect of Ptolemy's equant circles.) Since the eccentricity is three times the epicyclic radius, the planet is indeed at greatest distance from the Sun when at aphelion of the eccentric, but—and this was Maestlin's point—the epicycle extended farther yet from the Sun. If one was going to place a Copernican model like this inside one of the regular polyhedra, should the faces of the figure be made tangent to a sphere enclosing the path of the planet itself, or to the somewhat larger sphere enclosing the actual epicycle at its aphelion? Similarly, should the inscribed polyhedron touch the sphere at the planet's least distance, or the sphere at the epicycle's least distance? Figure 5.4 shows the alternatives for the outer sphere and the inner sphere.

These alternatives raised the now-unavoidable question of how seriously one should take the crystalline spheres of medieval cosmology. If there really were rigid, transparent spheres in the heavens, carrying the planets as they rotated, then the "sphere" of a planet in Kepler's polyhedral theory had to include room for all the spherical apparatus needed to move the planet. The polyhedron outside it would be larger, and the next planetary sphere circumscribed to that polyhedron

[10] Maestlin to Kepler, letters nos. 29, 37, 52, 58, and 63, all in *G.W.* 13. A. Grafton has analyzed Maestlin's assistance to Kepler at this juncture and has translated the relevant portions of the correspondence between the two of them, plus an appendix that Maestlin added to the *Mysterium* to explain the calculation of distances, in "Michael Maestlin's Account of Copernican Planetary Theory."

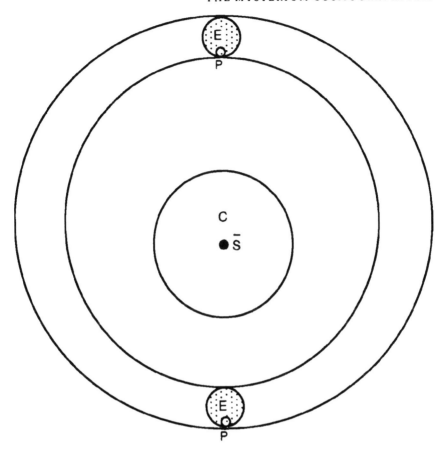

FIGURE 5.3. The Copernican epicycle

would be larger, and so on. Likewise, the polyhedron within the planetary sphere, and all the spheres and polyhedra within it, would be smaller. The planetary system as a whole would expand to cover a substantially greater range of distances. Maestlin originally calculated the minimum and maximum distances of the planets under just this assumption, evidently taking for granted that epicycles were a real part of the heavens.[11]

Kepler decisively rejected the idea of including rigid-sphere machinery in his figures. If that were done, the whole business would have to be abandoned, for the numbers were not even close. It took nothing from the nobility of his theory if the actual paths of the planets followed the proportions of the polyhedra, regardless of what spheres

[11] Grafton, "Michael Maestlin's Account," p. 531.

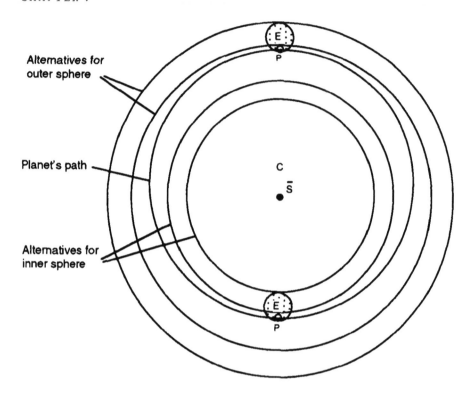

Alternatives for
outer sphere

Planet's path

Alternatives for
inner sphere

FIGURE 5.4. Leave room for the epicycle?

moved the planets. He did not himself believe in the reality of solid spheres, nor did Tycho.[12] The "spheres" among which he inserted his polyhedra were geometrical, not physical. They extended, he suspected, from the least to the greatest distance of the planet from the Sun. (He was not dogmatic about this but advised the reader to be guided by how well the numbers fit in interpreting this theory.[13])

Kepler presents the results of Maestlin's calculations in a table (table 5.2). The distances are given in degree notation, with the radius of Earth's orbit taken as 1°0′0″. Thus the distance of Saturn at its highest in column 3 is 9°42′, which is $9\frac{42}{60}$ times the radius of Earth's orbit. Column 3 gives distances from the mean sun—hence the highest and lowest distances for Earth are both 1, but the Sun has a certain distance from the center. Kepler is not interested in these distances. The Creator, he is certain, would have constructed not a structure embracing all the planets around the center of one planet's orbit but rather one

[12] G.W. 1.76.7–17.
[13] Ibid., 1.5–8.

TABLE 5.2.
Distance Comparison in the *Mysterium cosmographicum*

1	2	3	4	5	6
		° ′ ″	° ′ ″	° ′ ″	° ′ ″
Saturn	Highest	9 42 0	9 59 15	10 35 56	11 18 16
	Lowest	8 39 0	8 20 30	8 51 8	9 26 26
Jupiter	Highest	5 27 29	5 29 33	5 6 39	5 27 2
	Lowest	4 58 49	4 59 58	4 39 8	4 57 38
Mars	Highest	1 39 56	1 39 52	1 33 2	1 39 13
	Lowest	1 22 26	1 23 35	1 18 39	1 23 52
Earth	Highest	1 0 0	1 2 30	1 2 30	1 6 6
	Lowest	1 0 0	0 57 30	0 57 30	0 53 54
Venus	Highest	0 45 40	0 44 29	0 45 41	0 42 50
	Lowest	0 40 40	0 41 47	0 42 55	0 40 14
Mercury	Highest	0 29 24	0 29 19	0 30 21	0 28 27
	Lowest	0 18 2	0 14 0	0 14 0	0 13 7
Sun	Highest	0 2 30	0 0 0	0 0 0	0 0 0
	Lowest	0 1 56			

around the Sun itself. Column 4 gives the distances calculated from the actual Sun. Earth's distance from the Sun varies by 2½ parts out of 60 (Ptolemy's eccentricity, and Copernicus's maximum eccentricity). The Sun's distance from itself is of course zero.

For comparison, the solar distances implied by the polyhedral theory (in combination with the various planetary eccentricities) appear in columns 5 and 6. Column 5 contains a strict nesting of the planets, calculated by working outward from Earth. Earth's mean distance is by definition 1, so its eccentricity determines its greatest distance as about 1°2′30″. The dodecahedron (between Earth and Mars) has a ratio of 795:1,000 between its inscribed and circumscribed spheres. The inside of Mars's sphere is therefore at a distance of $1°2'30 \times \frac{1,000}{795}$, or about 1°18′39″. The ratio between the inside and the outside of Mars's sphere is determined by the eccentricity of that planet and is the same as the corresponding ratio in column 4. From that point the tetrahedron's ratio of spheres gives the distance to the inside of the sphere of Jupiter, Jupiter's eccentricity gives the distance to the outside of its sphere, and so on. Column 6 contains an alternate set of polyhedral distances based on the assumption that the sphere of the Earth had to be thick enough to include the orbit of the Moon between the icosahedron and the dodecahedron.[14]

The adequacy of the polyhedral model—with regard to the parameters of Copernican astronomy—is thus revealed in the comparison of column 3, the Copernican distances, with either column 5 or column 6. (Kepler has no a priori opinion on whether the Moon's orbit belongs within the sphere of the Earth.) The agreement is obviously far from perfect, but Kepler is not discouraged. The number of intervals between planets agrees perfectly with the number of perfect solids, and those intervals occur in the order that Kepler has persuaded himself is most natural to the perfect solids. The numbers agree reasonably well with Copernicus and might agree better if the true dimensions of the universe were known. Kepler does not believe that all this is accidental.

The Absence of Musical Harmony in the *Mysterium*

Along with the polyhedral hypothesis discussed above, the *Mysterium cosmographicum* contains Kepler's first attempts to understand the physical causes of planetary motion, and some passing remarks about the inadequacy of the Copernican version of heliocentrism—remarks that would guide his own rethinking of astronomy over the next decade. In chapter 20 he ponders a striking pattern that had long been assumed but that heliocentric astronomy has confirmed: that planets more distant from the Sun move more slowly than those closer to the Sun. He later developed his reflection on this fact into a penetrating physical analysis of planetary motion, an analysis that guided his study of Mars in the *Astronomia nova*. The relation eventually crystallized as his harmonic or "third" law of planetary motion. In chapter 22 of the *Mysterium*, he points out that the relation between motion and distance from the Sun holds even within the revolution of an individual planet. Near its aphelion, where it is farthest from the Sun, a

---

[14] This table contains many small errors, discussed in German by Caspar (*G.W.* 1.426–30) and in English by Aiton (Kepler, *The Secret of the Universe* [New York, 1981], pp. 244–46). The numbers in columns 3 and 4 had been calculated by Maestlin from Erasmus Reinhold's *Prutenic Tables*, as discussed by Grafton.

planet's motion is slowest, and near perihelion its motion is fastest. This insight, further developed in the *Astronomia nova* and the *Epitome*, emerged as the area law, Kepler's "second" law of planetary motion.

It was, in itself, a breakthrough merely for Kepler to regard a planet in this way, as a body moving through space, sometimes faster and sometimes slower. Astronomers had always relied on uniformly rotating spheres as the model of planetary motion. Such models had both physical and conceptual significance. The equant circle used by Ptolemy, for instance, had no physical existence; it intersected the "real" spheres in a way which ruled that out. As a mathematical circle, however, it served to measure the motion of the sphere that carried the planet in its epicycle. In the final chapters of the *Mysterium* Kepler cast off even this ghost of a sphere in his astronomy. Tycho had disproved the physical spheres; Kepler now abandoned their disembodied mathematical remnant.

It has frequently been remarked that this little book, which Kepler published in his midtwenties, indeed before he considered himself an astronomer or had much expertise in astronomy, contained already the seeds of his later work. There is a great deal of truth in this; one can find the first occurrence of almost any important theme from Kepler's work in the *Mysterium cosmographicum*. This is not so, however, for the musical harmonies that he sought (and found, eventually) in astronomy. Musical staves do appear in chapter 12, to illustrate that just as many harmonic intervals exist in music as regular solids in geometry (provided one counts the major and minor thirds as a single interval, and the two sixths as another); but the cosmological thesis of the book is based purely on geometry and not on music. The regular polyhedra, to be sure, are themselves the scaffolding from which he would later construct the detailed harmonic theory in the *Harmonice mundi*. It is not until the summer of 1599, however, that Kepler expresses a theory about the planetary motions that is specifically based on musical harmony.

Kepler's First Harmonic Planetary Theory

IN JULY OF 1599 Kepler wrote to one Edmund Bruce, an Englishman visiting Padua and in contact with Galileo, about a new theory that he hoped might attract the attention of the latter. A few weeks later he described the same theory in a long letter to the Bavarian chancellor J. G. Herwart von Hohenburg, a distinguished and enthusiastic amateur of astronomy and astrology and a frequent correspondent of Kepler's. Soon afterward he elaborated on it again to his former teacher, Michael Maestlin.[1] To all three of his correspondents he pointed out that the predictions of this new harmonic theory were better than those of the physical theory of motive souls that he had advanced in chapter 20 of the *Mysterium*. The latter theory has been widely recognized as the origin of Kepler's later physical theories of planetary motion; the more accurate harmonic theory has been overlooked by most historians.[2]

Suppose, Kepler suggests, that the planets moved through something like air. They would make harmonious sounds, as the strings of a musical instrument do when they move through the air. The chord would be that shown (in modern notation[3]) in figure 6.1.

These particular intervals would arise if the proportions of the six velocities, starting with Saturn, were as 3:4:8:10:12:16. In other words, equally long strings with tensions in the proportion 3:4 would make the interval of a fourth, as between Saturn and Jupiter; with tensions in

---

[1] Kepler to Edmund Bruce, letter no. 128 in *G.W.* 14.7–16; Kepler to J. G. Herwart von Hohenburg, letter no. 130 in ibid., 14.21–41. Kepler to M. Maestlin. letter no. 132 in ibid.. 14.43–59.

[2] Ibid., 14.13.257–59; 14.27.252–53, 14.28.280–82; and 14.51.353–52.394. Caspar discusses the harmonic theory, and Field describes it briefly, but neither goes into detail. In his letter to Maestlin Kepler also observed, correctly, that his physical theory from chap. 20 of the *Mysterium* could not be right, since it gave reasonable results only for adjacent planets. On this see Stephenson, *Kepler's Physical Astronomy* (New York, 1987). pp. 13–15.

[3] Kepler's musical notation was much the same, except that he was much freer in the placement of the clef signs on the staff. In this chapter I will, to avoid confusion, put clef signs in the places that have become standard and shift the positions of the notes on the staff to compensate.

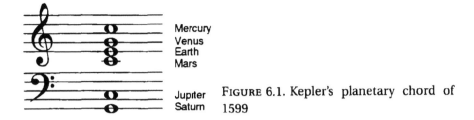

FIGURE 6.1. Kepler's planetary chord of 1599

the proportion 4:8 they would make the interval of an octave, as between Jupiter and Mars; and so forth. It seems at first incongruous, Kepler admits, that Saturn as the lowest or bass voice is only a fourth lower than Jupiter, instead of a fifth lower as is customary for the bass; but thus it is in the heavens.

Kepler obviously has chosen the six given values for the velocity (*vigor*) of each planet with attention to the harmonic ratios among them. The specified musical intervals correspond roughly, he remarks, to the spacing required by the regular polyhedra in his recently published *libellus*. He continues by calculating the sizes of the spheres that would revolve in the times observed for the planets, if they moved at the stated velocities, and comparing these sizes with those calculated from Copernican principles. The calculation is approximate, disregarding variations in velocity and treating each planet as if it moves uniformly on a circular orbit. Given the periodic times and (by his hypothesis) the proportional velocities, it is easy to estimate the circumferences of the circles.

As an example, the proposed velocity of Mars is to that of Earth as 8:10. In 365¼ days with velocity 10, Earth completes its orbit, having traveled a distance proportional to $10 \cdot 365\frac{1}{4} = 3,652\frac{1}{2}$. Mars, while completing its orbit in 687 days, travels a distance proportional to $8 \cdot 687 = 5,496$. The radii of the circles were in the same proportion as their circumferences, so this theory implied that the Martian orbit had a radius of

$$R_m = \frac{5,496}{3,652\frac{1}{2}} \cdot R_e$$

$$= 1,505, \text{ where } R_e = 1,000$$

Table 6.1 gives the results for all the planets, with the comparable "observed" distances (allegedly from the *Mysterium cosmographicum*) that Kepler reported to Bruce and Maestlin.

TABLE 6.1.
Orbital Radii from Harmonic Theory of 1599

1 Planet	2 Harmonic theory	3 Mysterium (to Bruce)	4 Mysterium (to Maestlin)
Saturn	8,837	9,164$\frac{1}{2}$	9,164
Jupiter	4,745	5,218	5,246
Mars	1,505	1,529	1,520
Earth	1,000	1,000	1,000
Venus	738	763	719
Mercury	385	369$\frac{1}{2}$	360

Kepler attributes the distances in column 3 to "Copernicus . . . , as computed in my little book"; they cannot, however, be derived from the *Mysterium* and are of mysterious origin. The somewhat different Copernican values reported to Maestlin a month later do agree with the values published in the *Mysterium*.[4]

The agreement of the harmonic theory with the distances implied by Copernican theory is far from exact; but it improves on the physical theory of the *Mysterium*, and the young Kepler is pleased. The theory allows him to calculate the planets' mean distances from harmonic principles. Since the polyhedral theory tells him the thickness of the empty spaces between the spheres, he thinks he can somehow distribute the space remaining among the thicknesses of the various spheres themselves—among the planetary eccentricities, in other words. He is uncertain just how this should be done and admits to both Herwart and Maestlin that he feels rather as if he has "a bird under a bucket."[5]

The harmonic progression that Kepler assigns to the planetary ve-

---

[4] The harmonic results of column 2 are from *G.W.* 14.13.236–41. The Copernican results in column 3 were reported to Bruce in ibid., 14.13.251–56, although Kepler mistakenly copied the harmonic value 738 for Venus. (He corrected his error in the letter to Herwart, ibid., 14.29.304.) The Copernican values in column 4, reported to Maestlin in ibid., 14.53.430–35, agree with the mean Copernican distances in the *Mysterium*, ibid., 1.74–75.

[5] ". . . Tanquam avem sub everso modio." Ibid., 14.29.308–9. ". . . Videor habere avem sub capaci modio." Ibid., 14.53.418.

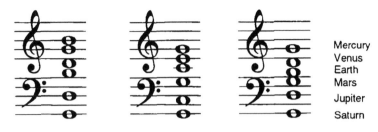

FIGURE 6.2. Alternate planetary chords of 1599

locities in this theory is constrained by his desire to obtain a pleasing chord. Among various plausible progressions, it yields distances much closer to Copernicus than do the others. To show this, he proposed to Maestlin three other harmonious chords (shown with modern notation in figure 6.2) that might be thought suitable for the planets.

To explain his choice between them, he tabulates the proportions and also the planetary distances that these imply if calculations like that shown above for Earth and Mars are carried out from the proportions. He includes in the table (table 6.2)[6] a variant form of the second chord (column 4) in which Jupiter is moved from c down to G and Saturn from G down to a low D. He also includes, for comparison, his preferred chord from figure 6.1 (column 6), with distances calculated to an additional significant digit.

The first chord moves the planets in the continued proportion 2:3:5:6:8:10. Assuming, as before, constant motion on a circle, these velocities and the well-known periodic times imply the distances given in column 2 of the table, where the Earth's distance is normed to 10,000. The proportions and distances for the other chords appear in the other columns, with the Copernican distances from the *Mysterium* in column 7. Clearly the distances from the preferred chord in column 6 fit better than those from any of the other chords, although not perfectly.

The chord in figure 6.1 was, therefore, the most accurate of the plausible harmonious chords. What is more, by the time he wrote Herwart and Maestlin in August, Kepler could show them an odd but ingenious way to derive from his arrangement of regular polyhedra three of the five intervals within the progression 3:4:8:10:12:16.[7]

---

[6] Ibid., 14.53. Caspar's corrections to a few of the numbers in this table are noted on 14.463.

[7] Ibid., 14.27.255–28.269; and 14.53.418–54.463. On 53.440 the symbols for Venus and Mercury should be Earth and Venus.

TABLE 6.2.
Orbital Radii from Other Possible Chords

1 Planet	2 First chord		3 Second chord		4 Variant of second		5 Third chord		6 Preferred chord		7 Copernican distances
Saturn	2	98,187	3	110,464	2	73,642	2	117,829	3	88,371	91,640
Jupiter	3	59,307	4	59,307	3	44,480	3	71,169	4	47,446	52,460
Mars	5	15,675	6	14,107	6	14,107	4	15,047	8	15,047	15,200
Earth	6	10,000	8	10,000	8	10,000	5	10,000	10	10,000	10,000
Venus	8	8,201	10	7,689	10	7,780	6	7,381	12	7,381	7,190
Mercury	10	4,016	12	3,614	12	3,614	8	3,855	16	3,855	3,600

The derivation is this: the cube separates the orbits of Saturn and Jupiter. Each solid angle of a cube comprises three plane angles of 90° each, for a total of 270° if all three are spread into a single plane. The ratio of 270° to a circle is 3:4, so it is fitting that the musical interval of a fourth, corresponding to the ratio of strings 3:4, should characterize the motions of Saturn and Jupiter, the planets whose spheres are separated by the cube.

Similarly, the tetrahedron separates the orbits of Jupiter and Mars. Each of the tetrahedron's solid angles includes three plane angles of 60°, making a total of 180°. The ratio of 180° to a circle is 1:2, and sure enough, an octave corresponding to the ratio 4:8 or 1:2 characterizes the motions of Jupiter and Mars in figure 6.1. The same pattern holds for the motions of Earth and Venus, which are characterized by the minor third and whose orbits are separated by the icosahedron. The five plane angles of 60° at each solid angle of an icosahedron amount to 5:6 of a circle, precisely the ratio of Earth's and Venus's minor third.

The pattern fails for Venus and Mercury. Their motions in this theory make the interval of a fourth, 12:16 or 3:4, whereas the four 60° plane angles making up an octahedron's solid angle total 2:3 of a circle. Kepler is forced to excuse this discrepancy by giving precedence to the polyhedra in determining the best ratio for these two planets. The octahedron of Venus and Mercury does not receive the proportion 3:4 of the plane angles making up its solid angle. Rather it receives the same proportion of motions as the cube, because the two figures are equally high (*aequealtum*). By this Kepler means that the octahedron fills out the same relative distance between its surrounding spheres as does the cube, since the proportions between the radii of circumscribed and inscribed spheres are equal for the octahedron and the cube. Because the octahedron is as high as the cube, it should correspond to the same musical interval as the cube, namely a fourth.

The pattern fails also for Mars and Earth, since three dodecahedral plane angles of 108° each make up 9:10 of a circle. For these planets Kepler appeals again to the dodecahedron's equal proportions of spheres with the icosahedron. Even after doing so, he must further adjust the minor third 5:6, borrowed from the icosahedron, into a major third, 4:5. Otherwise, he argues, there would occur a musical dissonance from the juxtaposition of two minor thirds.

Herwart's reaction to this early and imperfect harmonic theory was guarded. He was particularly concerned that the numbers did not agree very well and regarded the theory as a priori, based on a mere suspicion and not a demonstration; indeed he questioned whether it was capable of demonstration. Kepler's reply was revealing. Apparent numerical inaccuracy was a problem he had addressed in chapters 18 and 19 of the *Mysterium*, arguing that the *Prutenic Tables* (to which he compared his theories) were far from perfect. As for the accusation that it was an a priori theory, "First, I think that—aside from a few propositions—I have proposed, if not an ironclad demonstration, yet one which stands, in the absence of contrary argument. Second, the suspicion is not entirely false. For man is the image of the creator, and it may be that in certain matters pertaining to the adornment of the world the same things appear to man as to God."[8] Kepler believed, in other words, that his own sense of harmony reflected the Creator's preferences. It represented, therefore, an order that was objectively present in the universe.

After fondly describing to Maestlin his harmonic theory, its empirical superiority over his physical theory from the *Mysterium cosmographicum*, its possible use in deriving the planetary eccentricities, and its partial derivation from the regular polyhedra, Kepler speculated about how the great Tycho Brahe (whose establishment he had not yet joined) might receive it. He wondered whether Tycho's "little paper houses" (*domunculae chartaceae*) would be overthrown by his own celestial harmony and admitted his reluctance to contradict such a man.[9] In retrospect, it was not so much Kepler's harmonies as the things he discovered in his relentless pursuit of those harmonies that swept away all previous astronomy within a few years after this letter was written.

In 1599 Kepler still thought that he might be able to resolve the remaining uncertainties in his theories in short order. He hoped to receive from Tycho accurate values for the eccentricities of the planets. These could confirm the polyhedral theory he had advanced in the *Mysterium cosmographicum*. If he was able to justify or derive the eccentricities from harmonic principles, then his speculations would fuse into a single harmonic theory accounting for the structure of the heavenly world. Having realized the error of the physical theory advanced in the *Mysterium*, he had (temporarily) abandoned his physics[10] and looked at this time to harmonic speculation for insight

[8] Herwart to Kepler, 29 August 1599, no. 133 in *G.W.* 14.59.13–21; Kepler to Herwart, 14 September 1599, no. 134 in ibid., 14.73.426–45.

[9] Kepler to Maestlin, 29 August 1599, no. 132 in ibid., 14.54.468–71.

[10] "The law of the virtue emanating from the Sun is not simply that the weakening is proportional to distance." Ibid., 14.51.356–57.

into planetary astronomy. The full development of that speculation, requiring calculation of the eccentricities of planetary orbits from harmonic principles, turned out to be nearly a lifelong task. Kepler's ultimate solution, in book 5, chapter 9 of the *Harmonice mundi*, was vastly more complex than he could have imagined in 1599. That solution, remarkably enough, still followed the general outline of the proposals in the letters discussed in this chapter. The regular polyhedra gave the approximate sizes of the huge empty spaces separating the orbits. A simple, elegant pattern (here the harmonic sequence, later the "harmonic law") provided the mean distances as a function of the periodic times. The remaining space belonged to the eccentricities of the various planets and had to be somehow divided up among them. It was this last detail, the harmonic allocation of eccentricities, that took Kepler twenty years, off and on, to complete.

The Reconstruction of Ptolemy's *Harmonics*

KEPLER IN 1600 was actively seeking employment. His position as a Protestant mathematician, teaching at a Protestant school, in Catholic Styria, was becoming untenable under increasing pressure from the Counter-Reformation authorities. After a very anxious period, he joined Tycho Brahe at the beginning of 1601 in Prague. The two men had quite different expectations from their association. Kepler needed primarily to support his family; from Brahe he also hoped to obtain accurate values for the dimensions of the planetary system, which he could use to perfect his harmonic theory.[1] Tycho wanted to employ Kepler's skillful pen against his own (deceased) bitter enemy and predecessor Nicolai Baer, called Ursus in Latin, on many bitterly controverted matters, including Tycho's invention of the mixed geohelio-centric world system.[2]

Amid the stress of leaving Graz and negotiating his new position, Kepler received from Herwart a book that confirmed his belief in musical harmony as a structural principle of the heavens. This was the *Harmonics* of Ptolemy, a book that Kepler had long been eager to see. In August 1599, Herwart mentioned the opinion of Ptolemy on how many consonant intervals there were in music. Kepler quickly replied that if Ptolemy's book on harmony was not too large and burdensome for the messenger, he would very much like to see it. He asked again in each of his next two letters, in December and then in July 1600. In that same month, Herwart finally sent along the *Harmonics* in a Latin translation that had been published in 1562 by one Antoninus Gogavinus.[3] What Kepler read in it amazed him with its similarity to his own speculations, but he had no leisure to pursue the subject. He

---

[1] Kepler to Herwart , 12 July 1600, letter no. 168 in *G.W.* 14.130.102–131.145.

[2] Regarding Tycho's quarrel with Ursus (who died 15 August, 1600), see Jardine, *The Birth of History and Philosophy of Science* (Cambridge, 1984), chap. 1; Gingerich and Westman, "The Wittich Connection"; and Thoren, *The Lord of Uraniborg* (Cambridge, 1990), especially pp. 390–96, 432–40, 453–60.

[3] Letters no 133, *G.W.* 14.60.36–59; no. 134, ibid., 14.74.480–81; no. 148, ibid., 14.100.20–21; no. 168, ibid., 14.131.133–35; and no. 169, ibid., 14.137.35–43.

arrived in Prague with his family in October 1600, although it was quite a while before he was settled enough to work productively.

In October of the next year Tycho unexpectedly died of kidney problems, aggravated by his unwillingness to excuse himself from the table at a dinner where a great deal of wine was consumed.[4] With surprisingly little fuss, Tycho's office of imperial mathematician, with his instruments and observational records, passed to Kepler as his principal assistant.

As a result, Kepler's decision to accept Tycho's offer of employment had consequences different from what either of them expected. Tycho had assigned Kepler the task of deriving parameters for the model of Mars, a particularly recalcitrant problem, which had stymied others because the large eccentricity of Mars exposed the inadequacies of the model itself—the inadequacies, indeed, of all existing planetary models. After Tycho's death Kepler found himself with full access to the observations he needed to escalate his metaphorical struggle with Mars into something more resembling a scorched-earth campaign. He conducted it brilliantly (and ruthlessly) for half a decade. In the end the victory was his, but the ancient field of astronomy was left in ruins. Along the way, though, Kepler had found, or more often constructed, tools with which he was able to build a new astronomy. In 1607 he sent to press his account of the war, the *Astronomia nova*, on which much of his present-day fame rests.[5]

As his labors on Mars came to an end, Kepler received a letter that reawakened his interest in celestial harmony. In January of 1607, Herwart wrote that he was trying to obtain Ptolemy's *Harmonics* and similar works but had been unable to do so—having apparently forgotten that he had himself sent the book to Kepler more than six years earlier. Kepler reminded him of that, with some surprise, but expressed despair at ever making sense of the Latin rendition by Gogavinus. If Herwart could locate a copy in Greek of the Ptolemy, in particular, Kepler promised to reward him with a perspicuous commentary on it.[6]

---

[4] Thoren, *The Lord of Uraniborg*, pp. 468–69.

[5] Many accounts of the "war with Mars" exist; a recent favorite of mine is the voluminous third chapter of Stephenson, *Kepler's Physical Astronomy* (New York, 1987). The metaphor itself is from Kepler's dedicatory letter to Rudolph II in the *Astronomia nova* (*G.W.* 3.7–10).

[6] Herwart to Kepler, 16 January 1607, no. 407 in ibid., 15.375.15–20; Kepler to Herwart, late January 1607, no. 409 in ibid., 15.388.83–389.100.

In March Herwart was able to oblige. He sent Kepler a manuscript containing Ptolemy's *Harmonics* in Greek, with a commentary by Porphyry and another by "the monk Barlaam," who argued that the text attached to Ptolemy's last three chapter titles was inauthentic.[7] Kepler immediately used this Greek manuscript to confirm his poor opinion of the Latin translation by Gogavinus. He proposed to publish an edition of the text in Greek, with his own commentary and, perhaps, a new Latin translation, observing that although the book might not achieve success with the public, it surely belonged in libraries.[8]

Kepler never published his translation or his commentary. When the final book of the *Harmonice mundi* appeared in 1619, its table of contents promised an appendix containing a translation starting from chapter 3 of book 3 of Ptolemy's *Harmonics*, the chapters that treated the same material as Kepler's work. Kepler promised to complete Ptolemy's text, which was defective just at the critical part, in chapters 14 through 16, where Ptolemy had surely (Kepler thought) showed how to derive parameters of the planetary models from harmonic considerations. The appendix would also include commentary by Kepler, in which he planned first to explain and then to refute Ptolemy's theories and finally to compare them with his own.[9]

The actual appendix at the end of this same volume begins with a sheepish admission that the promised translation is not in the book. (The discrepancy between table of contents and appendix reminds us, if we need to be reminded, that setting type for so technical a work was neither easy nor swift.) Kepler began the translation ten years earlier, he explains, and had completed all of Ptolemy's work and half of Porphyry's commentary when his work was interrupted by his move from Prague to Linz, "combined with many other troubles." This was perhaps an understatement. Kepler's patron and employer, the Holy Roman Emperor Rudolph II, had been forced to abdicate in 1611 after

[7] Barlaam of Seminara (fl. fourteenth century), now chiefly remembered for trying unsuccessfully to teach Greek to Petrarch, was one of the first to doubt the authenticity of these final chapters. The text for Ptolemy's chap. 16 is now thought to be genuine, but Kepler accepted Barlaam's argument. On Barlaam, see R. Weiss, *Medieval and Humanist Greek* (Padua, 1977), especially pp. 11, 19, 38–39. On the authenticity of chaps. 14–16, see Barker, *Greek Musical Writings II* (Cambridge, 1989), p. 388 n. 86; and p. 390 n. 87.

[8] Herwart to Kepler, 6 March 1607, no. 412 in *G.W.* 15.408.9–23; Kepler to Herwart, April 1607, no. 424 in ibid., 15.449.1–451.77. See also Kepler to Herwart, 24 November 1607, no. 461 in ibid., 16.78–80.

[9] Ibid., 6.290.30–36.

the troops of his cousin Leopold entered Prague. Amid ensuing civic disorders, Kepler's wife Barbara had died suddenly of "spotted typhus." Rudolph himself died in January of 1612, and in April Kepler and his children departed Prague for Linz, to take up a position as district mathematician. There he had resumed his astronomical studies as best he could.[10]

It was only when he decided to publish his "five books of harmonies," the *Harmonice mundi* in other words, in 1617 or 1618, that Kepler saw an opportunity to compare his own discoveries in this field with those of his illustrious predecessor. He planned the appendix that was advertised in his table of contents: some thirty pages in which he would translate and complete the pertinent parts of Ptolemy's book 3, from chapter 3 to the end. He would compare Ptolemy's discoveries with his own and distinguish between Ptolemy's analogies and his own "legitimate demonstrations," making clear both the imperfection of those analogies and their origin in the false principles of geocentric astronomy. But war—the Thirty Years' War, no less—broke out in Bohemia. Travel was difficult, the available workers were all signed up as soldiers, and Kepler did not know when he would be able to get Ptolemy into press.[11]

The appendix that he finally managed to publish at the end of book 5 was much shorter, only six pages, and three of those were devoted to the esoteric theories of the English physician Robert Fludd.[12] Kepler's much longer Latin version of the *Harmonics* survived in manuscript, however, and Frisch finally included it when he published his edition of the *Harmonice mundi* in 1864.

We have already (in chapter 2) summarized those parts of Ptolemy's *Harmonics* that are relevant to our subject. Accordingly, we will now concentrate on what Kepler had to say about Ptolemy. Kepler's translation began with the third chapter of Ptolemy's book 3, which discussed the abstract nature of harmony. The next few chapters considered the nature of harmony within souls, a matter of great astrological interest. Kepler's notes were longer than Ptolemy's text; he had very definite beliefs about the nature of harmony itself, completely aside from its

---

[10] Ibid., 6.369.10–22. Caspar, *Kepler 1571–1630* (New York, 1962), pp. 210–16.

[11] Ibid., 6.369.33–370.10.

[12] Robert Fludd, *Utriusque Cosmi maioris scilicet et minoris metaphysica, physica atque technica historia* (Oppenheim, 1617–18). Kepler pointed out, correctly, that Fludd's interests, approach, and methods were utterly different from his own. Fludd took offense, and a pointless exchange of pamphlets ensued.

expression in the heavens. He thought it important, for example, to point out that harmony could not properly be said to exist in the absence of a rational soul, although Ptolemy's very brief text did not raise this question.[13]

From chapter 8 to the end of his book 3, Ptolemy proposed analogies between various aspects of musical harmony and particular motions in the heavens, as discussed in chapter 2 of this book. He compared his own version of the "perfect system," combining the GPS and the LPS, with the circle of signs or zodiac. Kepler regarded the perfect system as merely a convenient way of including all the different types of scale in one sequence, of no intrinsic importance in harmony. He found the analogy of the perfect system with the zodiac an impressive display of Ptolemy's ingenuity but unrelated to Kepler's own ideas about harmony. Ptolemy compared consonant intervals with planetary aspects, a comparison Kepler thought "in itself a thing of the greatest importance . . . , but that should be done very differently from the way Ptolemy did it here."[14]

Ptolemy turned at last to the planetary motions in chapter 10, although his geocentric conception of those motions precluded any close comparison to the harmonies Kepler perceived. Ptolemy's discussion was qualitative, based on broad analogies. He compared the planets' daily east-to-west motion "in length" to melody. He compared the motion of planets "in depth," from lowest to mean and highest, with the three genera of harmony in tetrachords. He compared motion "in width," north and south of the equator, with transitions in the *tonus*, or mode, of music. He compared the disposition of tetrachords within the perfect system with the planets' configurations relative to the Sun. Up to this point, Kepler admitted, there was little in Ptolemy that agreed with his own harmonic theories.[15]

The last three chapters, 14 through 16, seemed by their titles to bear directly on his subject matter. Unfortunately—or perhaps not, for Kepler's purposes—nothing authentic of these chapters remained except the titles; or so he believed. We know now that not all of the text for these chapters was supplied by Nikephoros Gregoras in the fourteenth century to fill the lacuna in the manuscripts. He wrote chapters 14 and 15 himself, but he also (correctly, it seems) moved the

[13] *O.O.* 5.336.

[14] Ibid., 5.364–70; 5.372, n. 1.

[15] Ibid., 5.378–89; *G.W.* 6.372.14–15.

text for chapter 16 from the middle of chapter 9, where it did not belong.[16]

Unabashed by the apparent absence of the three most important chapters for his purposes, Kepler restored them to explain what he thought Ptolemy must have meant and provided notes that corrected what he thought Ptolemy's errors would have been. One must remember that the reconstruction discussed below was that of Kepler, although he made a genuine attempt to base it on Ptolemaic astronomy and Ptolemaic harmony. Only in his notes did he comment on the errors of that astronomy. He had no comment at all on the musical principles he attributed to Ptolemy, although they were radically different from the ones he used in his own study of celestial harmonies.

These last three chapters were entitled as follows, in my translation of Kepler's translation:

14. Near which [relatively][17] prime numbers can the fixed notes of the perfect system be compared to the first spheres of those bodies that are in the heavens;

15. How the proportions of the proper motions can be expressed in numbers;

16. By what reasoning the families of planets can be compared with the families of sounds.

SIZES OF THE SPHERES

Ptolemy's chapter 14, judging from its title, had tried to establish that the fixed notes of the Perfect System could be found in the motions of the first, or outer,[18] spheres carrying the seven planets. (The two-octave Greater Perfect System used in Greek music theory was discussed in chapter 2. The seven fixed notes, at the ends of the tetra-

---

[16] It is now believed (see Barker, *Greek Musical Writings II*, pp. 388–90) that chap. 16 is genuine; it agrees, at any rate, with the end of Ptolemy's Canobic Inscription. Text for the disputed chapters, presumably that of Gregoras, had been published in the Latin translation of Gogavinus and was evidently present in Kepler's Greek manuscript. Kepler accepted the arguments of "the monk Barlaam" as disproving their authenticity and repeated some of them, which were based on fairly persuasive internal evidence. *O.O.* 5.410.

[17] Kepler added the qualifying "relatively" in a note. Ibid., 5.392, n. 2.

[18] Kepler remarked that "the first sphere of any planet is the outermost, including all the others necessary to the planet." Ibid., 5.392, n. 3.

chords making up the GPS, are shown in figure 2.4.) Ptolemy had probably continued to develop the analogy, Kepler speculated, by comparing the planets' lesser spheres with the movable notes of the Perfect System. These inner spheres varied the planet's distance from Earth, its *altitudo* as Kepler called it, according to the mathematical epicyclic model. Ptolemy in chapter 11 had already compared motion in depth—*kata bathos,* which Kepler had rendered as *in altitudinem*—to the changes in genus that musicians accomplished by altering the movable notes of the tetrachord. Kepler's speculation that chapter 14 elaborated on this comparison is perfectly reasonable and may be correct; we cannot know, since the text of the chapter is lost.

Kepler-as-Ptolemy begins by dealing with the fixed notes mentioned explicitly in the chapter title. He supplies a discussion that matches the seven fixed notes of the GPS neatly to the seven Ptolemaic planets (figure 7.1).

Saturn is at the bottom at proslambanomenos, the lowest note located a whole tone below the first tetrachord. Jupiter and Mars form the fixed notes of this first tetrachord, called hypaton. Hypaton literally means "the highest," as Kepler points out, so he thinks it entirely appropriate that this tetrachord should belong to the outer planets. Mars and the Sun form the next tetrachord, the meson, conjunct to the hypaton. Venus and Mercury form the tetrachord diezeugmenon, disjunct by a whole tone from the meson. Conjunct to this is the hyperbolaeon tetrachord, formed by Mercury and the Moon.

Mistakes in the Reconstruction

One cannot help noticing, at this point, that however appropriate figure 7.1 may be, it is very different from Ptolemy's authentic planetary music as shown in figure 2.6, which is based on the Canobic Inscription. Kepler does not realize, quite understandably, that Ptolemy placed Venus and Mercury on the same note; Ptolemy's reasons for doing this are equally mysterious to us. He does not realize that Ptolemy included both the fixed stars and the sphere of fire and air in his music, or that he included nete synemmenon and mese hyperbolaeon to make room for them. His biggest mistake, though, is to place the planets in the reverse of Ptolemy's order, giving the highest notes to the inner planets where Ptolemy gave them to the outer planets.

It is clear how Kepler made this mistake. As we shall see, he wanted

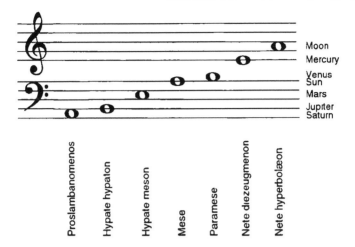

FIGURE 7.1. Kepler's reconstruction of Ptolemy's planetary music

to compare the "Pythagorean" music reported by Pliny with Ptolemy's music, to show the greater sophistication of the latter. Pliny's Pythagoreans had given high notes to high planets, Kepler believed, on the supposition that the higher planets moved much faster in order to complete their daily revolutions. Ptolemy, he was sure, was sophisticated enough to abstract the diurnal component from the planetary motions, leaving proper motions that are slower for the higher planets. Saturn, the highest, requires thirty years to circle the Earth if the daily rotation of all the heavens is ignored, and hence is the slowest planet and should receive the lowest note.

This argument is wrong but convincing. In the *Almagest* the diurnal rotation of the whole world is not a part of the individual planetary models, and Saturn's circle is indeed the slowest. It is hard to understand that which Kepler did not believe: that Ptolemy had given Saturn the highest note, corresponding to the fastest motion, among the planets. Yet there is the evidence, recorded from the stele at Canopus; Saturn is high and the Moon is low. Regardless of how one interprets the numbers from the Inscription, the ordering of the note names is clear, and nete hyperbolaeon is the highest note of any planet.

One way to understand why Ptolemy did this is to suppose that he was simply following a tradition, dating probably from the Pythagoreans, that naively placed the outer planets on higher notes because their motion, including the diurnal component, was faster than that of the inner planets. Another possibility, more intriguing, is to suspect that perhaps Ptolemy was not, when he wrote the *Harmonics*, as sophisticated an astronomer as Kepler thought him to be. The models of the *Almagest* are certainly mature Ptolemy; there is no particular reason to believe that the *Harmonics* is equally mature. Perhaps the Ptolemy who wrote down the planetary music of the *Harmonics* and the Canobic Inscription had not spent years dealing with planetary models from which the diurnal rotation had been abstracted. Perhaps he still had the naive point of view mentioned at the start of this paragraph. Perhaps he was following tradition because he had not yet developed the more sophisticated point of view credited to him by Kepler.

This is all speculative. At any rate, one cannot really blame Kepler for having failed to guess the fixed notes of the perfect system to which Ptolemy attached the motions of the planets. Let us now set aside our knowledge of what Ptolemy meant and consider Kepler's justification for the reconstruction depicted in figure 7.1.

Defense of the Reconstruction

All the planets' outer spheres, according to Kepler's reconstruction and figure 7.1, are on the outer or "fixed" notes of the tetrachords, as required by Ptolemy's chapter title. Saturn's proslambanomenos could be regarded as a fixed note, as noted in chapter 2, and Kepler regards it so—as, indeed, does Ptolemy. The remaining planetary spheres, which vary the planet's distance from Earth by moving it on an epicycle, and to a lesser extent by moving that epicycle along a circle eccentric to the Earth, correspond in Kepler's interpretation to the various intermediate or "movable" notes that can be inserted into the tetrachords between the fixed notes, to change between diatonic, chromatic, and enharmonic modes.[19]

In his notes Kepler points out how this musical diagram corresponds approximately to Ptolemaic astronomy. Among the superior planets, the musical notes for Jupiter and Saturn are separated by the smallest interval, only a tone; and indeed the outer spheres of those two planets are closest among the superior planets, since in Ptolemy, Saturn's epicycle is relatively small compared to the epicycles of Jupiter or Mars. The Sun and Mars are musically separated by a fourth, the largest interval among the superior planets; and astronomically the

---

[19] Ibid., 5.389–90.

distance between their outer spheres has to be large enough to leave room for the huge epicycle of Mars. Mars and Jupiter are likewise separated musically by a fourth. Although the proportional distance between their outer spheres is relatively much less than that between the spheres of the Sun and Mars, because of Jupiter's smaller epicycle, it is still quite large, Kepler argues, in absolute terms.[20]

Kepler defends similar analogies, with difficulty, for the inferior planets. Venus and the Sun have the smallest musical interval, a tone, and their outer spheres have the smallest astronomical distance, because the Sun lacks an epicycle. Mercury and Venus, on the other hand, have a large musical interval, a fourth, and their outer spheres have the largest distance in order to accommodate the big Venusian epicycle. The equally large musical interval between the Moon and Mercury is awkward, since the interval between the outer spheres of these planets is quite a bit less than that between the spheres of Mercury and Venus. Kepler (writing as Ptolemy) can only suggest weakly that this large musical interval corresponds to the astronomical distance between Mercury's outer sphere and the Moon's *inner* sphere, adjacent to the sublunary world and smallest of all the celestial spheres.[21]

For Kepler (as indeed for Ptolemy) a more exact numerical comparison between the musical and astronomical proportions is unavoidable. Reconstructing Ptolemy, Kepler expresses the fixed notes of the perfect system as lengths of strings that produce the correct notes when plucked. The intervals Venus-Sun and Jupiter-Saturn are whole tones (proportions of 8:9), and the others are perfect fourths (proportions of 3:4), so appropriate numbers for the musical intervals in figure 7.1 are those given in table 7.1,[22] which contains two tones and four fourths.

These numbers in column 2 are strikingly similar to those from the Canobic Inscription, but this is to be expected because both sets of numbers include all the fixed notes of the GPS. Ptolemy's numbers

---

[20] Ibid., 5.391. Table 4.1, from van Helden, *Measuring the Universe* (Chicago, 1985), gives a representative set of Ptolemaic distances. Those of interest are in column 4, for the outer spheres of the planets. The Jupiter-Saturn interval is proportionally the smallest among the superior planets. It is not the smallest absolutely, which vitiates Kepler's attempt to use absolute distances to explain the large musical interval between Mars and Jupiter.

[21] Ibid., 5.391.

[22] Ibid., 5.391.

TABLE 7.1.
Reconstructed Ptolemaic String Lengths

1	2	3
Moon	9	Nete hyperbolaeon
Mercury	12	Nete diezeugmenon
Venus	16	Paramese
Sun	18	Mese
Mars	24	Hypate meson
Jupiter	32	Hypate hypaton
Saturn	36	Proslambanomenos

(table 2.4) include the enigmatic $21\frac{1}{3}$, for the nete synemmenon from the LPS in his "complete" system, which is not present among Kepler's numbers. The association of planets with numbers in table 7.1 is, of course, in the reverse order of Ptolemy's.

Musically, the notes in table 7.1 span a range of two octaves, because Saturn's string has a length quadruple that of the Moon. The corresponding span of distances is surely much greater, Kepler knows, although he can only estimate how large Ptolemy thought it to be. He estimates as follows: the Sun, which is the midpoint of the musical system, is surely the geometrical midpoint of the astronomical system. The best Ptolemaic estimate of the proportion of the distance of the Moon to that of the Sun is about 1:19.[23] If the proportion of the distance of the Sun to that of Saturn equals this, then the overall span of distances would be twice 1:19, or 1:361. On the other hand, the best Ptolemaic estimate of the Sun:Saturn proportion is about 1:16. If one accepts this for the Moon:Sun proportion, then the overall system exhibits a span of distances of 1:16 compounded twice, or 1:256.

Kepler, writing as Ptolemy, describes this proportion of 1:256 in a most unusual way: "The outer sphere of Saturn equals the inner sphere of the Moon quadrupled not merely once but fully three times [*non semel tantum, sed plane ter quadrupla*], that is, 256 times as large

[23] Ptolemy, *Almagest*, book 5, chap. 15; pp. 255–57 in the Toomer translation.

and more." This is simply wrong. One must take a ratio of 1:4 not merely three times but four to obtain a ratio of 1:256. Kepler may have slipped in referring to a ratio of 256 as a quadruple ratio taken three times, but he uses this terminology consistently in his reconstructed chapter 14. He supposes, for instance, that Ptolemy preferred to hold the span of astronomical distances to 1:256, "since in this way the proportions of the heavens are made commensurable with the proportions of the musical notes, namely three times as large [*triplae scilicet earum*]."[24]

At any rate, Kepler-as-Ptolemy proposes that the proportions of the heavenly spheres should be "three times" (meaning four times) as great as the proportions of the strings yielding the musical notes in figure 7.1. From Saturn to Jupiter is a whole tone, 8:9 musically, so he obtains the proportion of their outer spheres by squaring 8:9 to get 64:81, and then squaring that to get 4096:6561. This is likewise the proportion of the spheres of the Sun and Venus, since a tone separates those planets also. The other three proportions equal a fourth, 3:4, squared and then squared again, or 81:256. Starting, therefore, with 81 and 256 as the proportional distances to the Moon and Mercury, Kepler compares the harmonic distances to those that (Ptolemaic) astronomy gives, expressed in semidiameters of the Earth.[25] Coincidentally, the astronomical estimates in terrestrial radii are on a scale rather similar to the numbers Kepler obtains simply by starting with the abstract terms of the proportion 81:256. Table 7.2 shows his comparison of the astronomical distances with the harmonic distances he had reconstructed for Ptolemy.[26]

The harmonic distances in column 3 supposedly arise from repeated multiplication by the quadrupled musical ratios $(\frac{4}{3})^4 = \frac{256}{81}$ or $(\frac{9}{8})^4 = \frac{6,561}{4,096}$, using the former ratio for planets separated by a fourth, and the latter ratio for planets separated by only a tone. The numbers for Venus, the Sun, and Saturn are not calculated accurately, however, indicating that this restoration of Ptolemy's final chapters was left in draft form. Kepler based the astronomical distances in column 4 on the traditional assumption that the spheres carrying the planets are as thin as possible, given the smaller spheres they must contain, and that

---

[24] *O.O.* 5.391.

[25] Ibid., 5.392. On line 3, the number 9 is omitted from the text; read *ut proslambanomenos 9 ad hypaten hypaton 8 ter.*

[26] Ibid., 5.392.

TABLE 7.2.
Kepler's "Ptolemaic" Distances

1 Planet	2 Interval	3 Harmonic distance	4 Astronomical distance
Moon		81	64
Mercury	fourth	256	167
Venus	fourth	817	1,070
Sun	tone	1,312	1,176
Mars	fourth	4,147	8,022
Jupiter	fourth	13,107	12,420
Saturn	tone	20,493	18,094

they are nested without wasting any space between them. The origin of these exact "Ptolemaic" distances is slightly mysterious, although they closely resemble the distances given by al-Battānī ("Albategnius"), whose planetary distances Kepler cited in a later note to this chapter.[27]

In notes appended to his reconstruction of the text, Kepler points out the defects of this hypothetical theory—starting, of course, with its faulty astronomical premises and its poor agreement with the numbers derived from those premises. A more subtle problem, one reflecting Kepler's honest desire to assess Ptolemy's theory in the context of Ptolemaic music and astronomy, arises from his unawareness that his predecessor had assigned higher notes to the outer planets. As discussed above, Kepler assumes that Ptolemy, as a sophisticated astronomer, must surely have abstracted planetary motions from the diurnal rotation of the heavens and must therefore have considered Saturn the slowest-moving of the planets, as in the planetary theory of the *Almagest*.

---

[27] Van Helden, *Measuring the Universe* (Chicago, 1985), p. 32, where the greatest distances for Venus, Mars, and Saturn agree exactly with these. Those for the Sun and Jupiter are close, whereas the Moon and Mercury are extremely close—within 1. *O.O.* 5.395, n. 13.

If this is so, Kepler reasons, Ptolemy would have given Saturn the lowest of the fixed notes, proslambanomenos. The second analogy, which Kepler has plausibly supplied, of a planet's inner spheres with moveable notes, leaves Saturn bare, for no moveable notes are associated with proslambanomenos. This, Kepler thinks, would have been a real problem for Ptolemy, who in his astronomy gave Saturn an epicycle and an eccentric fully analogous to those of the other planets.[28]

Pythagorean Harmonies

Kepler deduces, from the chapter heading of Ptolemy's fourteenth chapter, that Ptolemy was himself describing, and trying to improve, the Pythagorean teachings about the music of the heavens. Kepler's source for the latter was Pliny, who (as we have seen) asserted that Pythagoras compared planetary intervals with musical intervals. As the eight notes in the octave defined seven musical intervals, so the eight bodies making up the heavens were separated by seven distances or "intervals," encompassing a range of seven whole tones. Hence the analogy in Pythagoras was of heavenly bodies to sounds, and of intervals of heavenly bodies to intervals of sounds.

As an aside, Kepler's discussion of Pythagorean planetary harmonies presupposes a geocentric arrangement of the planets. Like other early Copernicans seeking the approbation of antiquity, Kepler generally liked to ascribe a heliocentric cosmology to "the Pythagoreans," on the basis of Aristotle's account in *de Caelo*. He never explicitly attributed heliocentrism to Pythagoras himself, and it is a delicate question how far Kepler distinguished between Pythagoras and his followers. At any rate, the "Pythagorean" heavenly spheres Kepler discusses in his notes to Ptolemy's *Harmonics* are geocentric spheres.

Regarding that theory, he realizes that Pythagoras had been misinterpreted by Pliny, who divided the octave among the seven intervals separating eight heavenly spheres, and amounting in sum to seven whole tones. Kepler points out that *six* whole tones suffice to make up an octave, and concludes that the original theory must surely have placed the octave between the Moon and the zodiac, omitting the interval from Earth to Moon. He surmises that the ambiguity in the word *interval* fooled either Pythagoras or his interpreter, Pliny. There is certainly a geometrical interval (a distance) between the Earth and

[28] *O.O.* 5.394–95, n. 12.

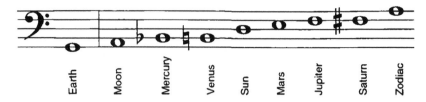

FIGURE 7.2. Kepler's reconstruction of Pliny's Pythagorean planetary music

the Moon, but there should be no musical interval between them because Earth, which does not move, cannot be associated with any sound. Bodies at rest make no sound.[29]

Lower planets—planets closer to Earth—are associated with lower notes in the Pythagorean harmonics, unlike Kepler's Ptolemaic harmonics (or his own). That ordering was appropriate, Kepler acknowledges, if Pythagoras thought of the heavenly sounds as produced by the diurnal motion of the spheres. Since each sphere rotates westward in twenty-four hours, in addition to its much slower movement eastward, the higher and larger spheres move faster overall and produce higher musical notes, according to this viewpoint.

Kepler transcribes the Pythagorean theory of planetary harmony, as he reconstructs it from Pliny, as in figure 7.2. From Earth to Moon is a whole tone, outside the octave of the moving bodies, as Kepler thinks it should be. The remaining intervals are a mixture of tones, semitones, and tones-and-a-half as specified in Pliny, although Kepler does not bother to explain their arrangement.[30] This is a perfectly good realization of the Pythagorean scale reported by Pliny. (We gave no comparable figure for our account of Pliny in chapter 2, since Pliny—unlike Kepler—did not indicate particular notes for the planets.)

Comparing his reconstructions of Pythagoras and of Ptolemy, Kepler finds each to have its good points. Both err in the ordering of planets, for indeed geocentric astronomy can provide no sound basis for determining that order. Ptolemy distinguishes the intervals between the spheres from the motions of the spheres, postponing a harmonic analysis of planetary motions until chapter 15. Pythagoras, on the other hand, treats these as a single question. The reason is clear to Kepler: to Pythagoras a planet's motion is at least in large part the diurnal rotation, about which little can be said except that it moves

---

[29] Ibid., 5.395–96.

[30] The above account is from ibid., 5.395–96, n. 13. After giving this scale according to Pliny, Kepler remarks that Macrobius also explains it a little less suitably.

larger spheres faster, absolutely, than it moves smaller spheres. Ptolemy, in Kepler's opinion, abstracts the diurnal rotation, treating it as a kind of rest, a reference point to which individual planetary motions could be compared.

This supposed difference explained why Pythagoras gave faster motions, and higher notes, to the outer planets, while Kepler is sure that Ptolemy considered them to be slower than the inner planets. Indeed, Ptolemy's separate treatment of planetary motions in chapter 15 proves (to Kepler) that he was concerned with motions that differ among the planets. These can only be the proper motions, after the removal of the diurnal rotation. The proper motions are slowest for the highest planets. He is sure, therefore, that his reconstruction was "not alien to the mind of Ptolemy" when it assigned the lowest note to Saturn and the highest to the Moon.[31] (As we argued in chapter 2, Kepler was probably wrong. The Ptolemy of the Canobic Inscription, and probably also of the *Harmonics*, did not make this abstraction but rather retained the more naive point of view that Kepler here attributed to Pythagoras.)

Ptolemy's separate treatments of the sizes and motions of planetary spheres, although it supposedly arose from a more sophisticated conception of planetary motion, apparently led him (in this reconstruction of his fourteenth chapter) to compare sizes with musical notes. There can be no sound without motion, however, and in that sense Kepler finds the simpler Pythagorean theory to be more reasonable. He ends his notes on chapter 14, curiously enough, by finding the Pythagorean theory "altogether more elegant and richer in mysteries"[32] than the Ptolemaic.

Having compared his two illustrious reconstructed predecessors, Kepler cannot resist comparing his own harmonic discoveries with those he ascribes to Ptolemy. He praises Ptolemy's inspired realization that there is an archetype from which not only the number but the sizes of the heavenly spheres was determined. This "divine axiom" he shares; beyond that he cannot concur. The astronomy of Copernicus has provided Kepler with certain knowledge of the number, order, and sizes of the planetary spheres, even as his own "truer discipline" of astronomy indicates that the spheres are not solid but merely spherical regions of space. The epicycles that cluttered the Ptolemaic system are gone, and the basic facts of cosmology—most of which were con-

[31] Ibid., 5.397.
[32] Ibid., 5.398.

jectural in the old astronomy—are for Kepler established on a scientific basis.

Regarding the archetype itself underlying the heavens, Ptolemy discovered the existence of a harmonic order (*politia*) yet failed to discern its origin, which is properly to be sought in geometry and not in arithmetic. He combined into one harmonic system three things that ought to be kept distinct: the configurations, or astrological aspects; the sizes of the celestial spheres; and their motions. Kepler himself believes that the ultimate origin of all harmony is to be sought in geometry, and by pursuing this belief he has come to see the five perfect solid figures as the archetype for the heavenly spheres.[33]

MOTIONS OF THE SPHERES

Kepler thinks that Ptolemy, in his missing fifteenth chapter, probably discussed the "proper" motions of planets, that is, their motions abstracted from the diurnal rotation of the heavens. The chapter title does not indicate any comparison with harmony, and indeed all the traditional divisions of harmony have been treated in earlier chapters; yet Kepler assumes that Ptolemy tried to connect these motions to harmonic concepts. The connection is relatively easy in a heliocentric system, as no one knows better than Kepler. With Ptolemy's geocentric models, however, the great synodic epicycles confuse the picture greatly by making the planets slow down and stop, reverse direction, then slow again and resume their former direction. As Kepler notes, this extremely peculiar behavior renders any comparison, no matter how it is done, "inept and abhorrent." He conjectures that Ptolemy was forced to treat separately the motions of the eccentrics, the motions of the synodic epicycles, and the composite motions as seen from Earth.[34]

The various eccentrics have their respective mean motions, among which Ptolemy could certainly have calculated proportions. He might have been hard-pressed, Kepler remarks, to explain why three of the mean motions are the same. Kepler further notes that the eccentric

[33] For details he referred the reader to specific pages in the *Mysterium cosmographicum* (published "22 years ago," i.e., in 1596) and in the *Harmonice mundi*, further evidence that Kepler was still working on the translation of Ptolemy in 1618, while the *Harmonice mundi* was in press.

[34] *O.O.* 5.404, n. 4; 5.400.

spheres are transparent, which is to say invisible. The mean period of a planet's return measures the motion of its eccentric sphere, but not of anything that can be perceived; and to Kepler's mind the whole point of harmony is that it be perceived. In any event, he himself has shown in the *Harmonice mundi* that the proportions among the planets' periods are not harmonic.[35]

Kepler next contrives for Ptolemy an ingenious investigation into the harmonic content of epicyclic motion. He attributes the diatonic genus of music to Saturn, Jupiter, and Mercury, planets that have small epicycles rotating more quickly than their eccentrics; the enharmonic genus to Mars and Venus, which have large epicycles rotating more slowly than their eccentrics; and the chromatic genus to the Sun and Moon, the luminaries, which embarrassingly lacked synodic epicycles altogether.[36] The motions of the epicycles, like those of the eccentrics, are in no way observable; in Kepler's own opinion the invisibility of these harmonies, to anyone or anything, is a grave drawback to this (or indeed any) Ptolemaic theory of celestial harmony.

The observed, composite motions in Ptolemy's system are still harder to harmonize. Fast, slow, stationary, even retrograde, they exhibit all possible harmonic (and nonharmonic, as well) proportions, and hence they can hardly be said to embody any of these proportions. Kepler does as well as he can, supplying a text that recites the apparent motions at epicyclic apogee and calculates the greatest and least proportions of these to the motions of the luminaries. No strong conclusions are to be had from any of this. To salvage some musical content, Kepler's Ptolemy makes a clever analogy between the composite motions of the planets, backward and forward in the heavens, and the rhythmic oscillations of the chorus on the stage of ancient Greek drama. After all, Kepler remarks, the most notable fact in all astronomy—a fact whose origin was still mysterious to Ptolemy—is that the forward-and-backward dance of the planets is closely tied to their position with respect to the Sun.[37]

Commenting on the clumsiness and indeterminacy of all this, Kep-

[35] Ibid., 5.404–5, n. 6. Kepler's demonstration of this is in book 5, chap. 4, but this note incorrectly cites chap. 6. Evidently he had drafted book 5 before annotating Ptolemy's *Harmonics* but had not revised it into its final form. The notes on Ptolemy were surely written before Kepler discovered his third law; afterward he was sufficiently occupied with his own theories.

[36] The Moon's epicycle rotated in the opposite direction from its eccentric and had an altogether different purpose from the synodic planetary epicycles.

[37] *O.O.* 5.400–2; 5.407, n. 11.

ler is emboldened to suggest that the lacuna at the end of Ptolemy's *Harmonics* might not be entirely without a higher purpose. Just as, elsewhere in astronomy, eclipses of the Sun and Moon are not reckoned as harmful, since they light the way to a knowledge of the heavenly motions, perhaps divine Providence has arranged for this defective text to lead readers—already enraptured, no doubt, by the glory of Ptolemy's vision—to Kepler's truer heliocentric expression of that vision.[38]

FAMILY RELATIONSHIPS

Ptolemy's chapter 16 is astrological, explaining how the affinities and antipathies of musical notes are analogous to those of the planets. Kepler has the text as restored by Nikephoros Gregoras (see n. 16) but rejects it, ironically, because it contradicts his firm belief that Ptolemy gave lower notes to the outer planets than to the inner planets.

His attempt to reconstruct it is halfhearted. Astrologers recognized dozens of reasons why heavenly bodies might be hostile or friendly to one another. Kepler, whose astrological beliefs are idiosyncratic, has little regard for these analogies and little enthusiasm for concocting an account of them. Fortunately Macrobius included bits of Ptolemy's *Harmonics* in his commentary on the *Somnium Scipionis*.[39] Kepler uses it—at times nearly word for word, as he admits—in reconstructing what Ptolemy might have said about the related subject. He actually suspects that Macrobius used the same, supposedly corrupt, text for chapter 16, in which case it must be very old indeed.[40]

In Kepler's notes he dismisses this astrological material and states firmly that these parts of astrology—an inescapably geocentric science—have nothing to do with the true harmonies of (heliocentric) planetary motion. He summarizes his own theories about the qualities of the planets from the astrological tract *Tertius Interveniens*, which he published in 1610. The theories are physically based and depend on the "hot" radiation of the Sun and the "humid" radiation of the Moon. After this personal digression he closes his commentary with a disarming caution to his reader: "This is my game; but it is metaphysical

---

[38] Ibid., 5.405, n. 7.

[39] Macrobius, *Commentary on the Dream of Scipio*, 1.19; in the Stahl translation (New York, 1952), pp. 167–68.

[40] *O.O.* 5.412.

or physical, having nothing to do with the harmonies of the planets. Use it if you wish; correct it if you can."[41]

PTOLEMY'S INFLUENCE ON KEPLER'S PLANETARY HARMONIES

It might appear that Kepler was scarcely justified in his frequent praise of Ptolemy's *Harmonics* as an inspiration for his own work on the subject. The first thirteen chapters of Ptolemy's book 3 have very little to do with astronomy, and that little consists of the broadest sort of analogy. Their influence on Kepler can only be characterized as inspirational. His personal belief that harmony pervades the world was not unique in his time, but it was more highly developed than was common, and he quite plainly found support in Ptolemy's similar attitude.

The last three chapters of the *Harmonics*, whose titles Kepler cherished, had no extant text at all that he trusted; and even his reconstruction of that text was far from impressive as a demonstration of harmonic structure in the world. Time and again Kepler's reconstructed text stumbled on the inadequacy of Ptolemaic astronomy—first in being geocentric; also in relying on a physical machinery of rotating spheres that are not perceptible and hence not suited to the expression of harmony, for harmony is a phenomenon that cannot exist without perception. Kepler cannot imagine what sort of soul might appreciate harmonies among invisible epicycles and eccentrics. The perceptible motions of Ptolemaic astronomy, on the other hand, are composite and variable. Comparison of their extreme values is thwarted by the retrogradations that regularly reduce them to zero and then beyond.

Kepler had a hard time explaining how Ptolemy could have elicited from this material a theory of celestial harmony that was at all convincing. And indeed, he lost much of his enthusiasm for Ptolemy's *Harmonics*, and gave up his plans to publish it, after he discovered the relation between periodic times and orbital radii known as his "third law." With that discovery Kepler's pursuit of harmony in the heavens ceased to resemble Ptolemy's and became something far more impressive: the fifth book of the *Harmonice mundi*.

[41] Ibid., 5.411–12, n. 5.

The *Harmonice mundi*

THE *HARMONICES MUNDI LIBRI V*, to which Kepler had intended to append his translation of Ptolemy's *Harmonics*, appeared finally in 1619. The principal theses of its five books were that certain ratios, arising from the eternal geometry of regular polygons, were particularly noble; that the influence of music on the human soul depended upon these ratios, as did the influence of astrological aspects on mundane matters such as the weather and the human soul; and that these same ratios had been systematically embodied in the creation of the solar system. Book 5, with which we shall primarily be concerned, sought—and found—harmony in the details of planetary astronomy.

Kepler did not reveal the full complexity of his harmonically-derived cosmology until very close to the end. The ninth of book 5's ten chapters explained, in near-obsessive detail, how the planetary system was as harmonious as it possibly could have been. The whole difficulty in showing this was to determine what had been possible. Harmonies were mathematical ratios and subject to the eternal laws of mathematics. Not even God could have constructed certain proportions in the heavens without having certain other proportions appear as a necessary consequence. Chapter 9 tried to show, among other things, that these internal constraints were the principal reason why the harmonic structure of the world was less than perfect.

MUSIC THEORY

To understand the musical harmonies that Kepler discovered in the heavens one must know something of music theory as it had been developed by the end of the sixteenth century.

Kepler was a largely self-taught but rather learned music theorist, who gave his most complete account of his own theory of musical harmony in book 3 of the *Harmonice mundi*. He adopted the system of "just" intonation instead of the older "Pythagorean" intonation. Pythagorean harmony was based on ratios among the numbers 1, 2, and

3. The intervals recognized as harmonic are the octave, corresponding to the proportion 1:2; the fifth, corresponding to 2:3; and the fourth, corresponding to 3:4, which is the difference between the octave and the fifth. In the more recent system of just intonation, major and minor thirds and sixths are accepted as consonances and assigned proportions that please the ear. Kepler developed a geometric explanation for why there should be precisely seven basic consonances, but it is clear that he accepts the thirds and sixths on grounds that are empirical and aesthetic: he likes the way they sound.[1]

With pleasant-sounding thirds and sixths available, the system of just intonation supports polyphonic music, music with multiple voices sounding simultaneously. Such music had become popular in the sixteenth century, and Kepler hails it as a fundamental improvement over the music of antiquity. (He believes that the harmony of the Pythagoreans, and of ancient Greece in general, was essentially monophonic, arising from a single voice.) Kepler attributes the failure of the theories Ptolemy advanced at the end of his *Harmonics* to the absence of polyphony in his harmonies, no less than to the inadequacy of his geocentric astronomy.

Table 8.1 shows the important musical intervals and the proportions to which they correspond. Kepler divides harmonious intervals into the *consonantia*, intervals that sound harmonious together, and the *concinna dissonantia*, or simply *concinna*, intervals that are agreeable when used in melody but not when sung together. The first four *concinna*, from the major tone to the diesis, are differences between consonances, whereas the two limmas and the comma (less pleasing intervals, whose status as true *concinna* Kepler doubts) could be obtained as differences between the larger *concinna*. Note that the diesis is smaller than either limma.

Neither the Pythagorean system nor Kepler's system of just intonation divides the octave into twelve equal semitones. The smallest intervals in Kepler's octave are variously semitones, limmas, and dieses, as shown in table 8.2.[2] This table, adapted from Kepler's book 3, chapter 8, shows all the notes in an octave; the lengths of strings that, if

---

[1] Walker, "Kepler's Celestial Music," in *Studies in Musical Science in the Late Renaissance* (London, 1978), p. 47. Walker shows that Kepler's geometric theory could have accounted for the Pythagorean intonation much more naturally and argues that his preference for just intonation must therefore have been genuinely empirical.

[2] *G.W.* 6.154–55.

TABLE 8.1.
Principal Intervals Used by Kepler

1 *Consonantia*		2 *Concinna*	
Octave	1:2	Major tone	8:9
Major sixth	3:5	Minor tone	9:10
Minor sixth	5:8	Semitone	15:16
Fifth	2:3	Diesis	24:25
Fourth	3:4	*Doubtful concinna*	
Major third	4:5	Limma	128:135
Minor third	5:6	Platonic limma	243:256
		Comma	80:81

plucked, yield the correct proportions among the notes; and the names of the intervals between adjacent notes.

The Types of Scales and Chords

Kepler employs musical scales and chords of two types: durus (literally "hard") and mollis ("soft"). They correspond roughly to the modern concepts of major and minor tonality, as these would be developed during the seventeenth century. The exact way in which Kepler uses these terms, however, is apparently unique. Dickreiter argues that Kepler's genera are conceived not harmonically but melodically, as in medieval church music.[3] An enthusiastic advocate of polyphonic harmony, in music and in cosmology, Kepler still does not have a fully polyphonic theory of harmony. His analysis in book 5 of the great chords played by all the planets together is based on a transitional, and incomplete, theory of the two types of harmony.

The essential feature of Kepler's genus durum is the presence of the

[3] Dickreiter, *Der Musiktheoretiker Johannes Kepler* (Bern and Munich, 1973), especially pp. 160–70.

TABLE 8.2.

Kepler's System of the Octave

0	1	2	3	4	5	6	7	8	9	10	11	12
2,160	2,048	1,920	1,800	1,728	1,620	1,536	1,440	1,350	1,296	1,215	1,152	1,080
G	G♯	A	B♭	B	C	C♯	D	D♯	E	F	F♯	G

Limma		Semitone		Semitone		Semitone		Diesis		Limma		
	Semitone		Diesis		Limma		Semitone		Semitone		Semitone	

major third and the major sixth in a scale, and the presence of either (or both) in a chord. The genus molle is similarly characterized by the minor third and sixth. Kepler follows contemporary practice in taking the bass note as G in his scales and chords, so a hard or durus scale is one that includes B♮ and E♮, and a soft or mollis scale one that includes B♭ and E♭.[4] The choice between F or F♯ as leading tone is not critical to the distinction between types; Kepler generally uses F♯ in the genus durum but allows either in the genus molle. Table 8.3 shows the notes and the types of tones making up each type of scale. It is easy to verify in this table that a major tone is composed of a semitone and a limma, whereas a minor tone is composed of a semitone and a diesis.

Kepler's division of chords into the genera differs more from modern concepts of tonality than does his division of scales. He does not recognize inversions as variant forms of the same chord, but instead characterizes any chord by its actual lowest note. This causes no problems with what we know as the standard G-major and G-minor triads (figure 8.1a).

The major and minor triads are respectively durus and mollis because of the notes B and B♭, which are the durus and mollis thirds in the scale from G. The C-major and C-minor triads, inverted to put G in the bass (figure 8.1b) are likewise durus and mollis respectively for Kepler because of the notes E and E♭,[5] the durus and mollis sixths in

---

[4] In discussing music, I see no benefit from retaining the traditional German notation for the two notes between A and C. In accordance with English-language custom, I always write B♭ where Kepler wrote B, and B where Kepler wrote H. As discussed below (regarding the skeletons of universal harmony in book 5, chap. 7), Kepler regards the ♭ symbol as a way of writing B, that is, B♭, and never uses it in text. Thus he always writes D♯ for E♭, and so on.

[5] Kepler writes E♭ on the staff but D♯ in text, as noted in n. 4.

121

TABLE 8.3.
Kepler's Scales

0	1	2	3	4	5	6	7	8	9	10	11	12
2,160	2,048	1,920	1,800	1,728	1,620	1,536	1,440	1,350	1,296	1,215	1,152	1,080
G	G♯	A	B♭	B	C	C♯	D	D♯	E	F	F♯	G

Limma		Semitone		Semitone		Semitone		Diesis		Limma		
	Semitone		Diesis		Limma		Semitone		Semitone		Semitone	

Durus Scale

G		A		B	C		D		E		F♯	G
Major tone				Semitone			Minor tone				Semitone	
	Minor tone				Major tone				Major tone			

Mollis Scale

G		A	B♭		C		D	D♯		F		G
Major tone			Minor tone				Semitone			Major tone		
	Semitone				Major tone				Minor tone			

the scale from G. However, the E♭-major and E-minor triads, inverted to put G in the bass (figure 8.1c) are respectively mollis and durus, despite their tonality in later harmonic theory, because B♭ and E♭ are the mollis third and sixth from G, while B and E are the durus third and sixth from G. This configuration prevents one from simply reading durus and mollis as major and minor (which are, after all, *Dur* and *Moll* in German).

In harmonies with more than three notes, such as the great six-part harmony of all the planets that we will discuss later, Kepler determines the genus in the same way. Placing the lowest note, corresponding most often to the aphelial motion of Saturn, on G, he considers chords containing B or E to be durus, and chords containing B♭ or E♭ to be mollis.

NOTATION AND USE OF PROPORTIONS

Mathematical proportions are for Kepler the language in which the world's harmony was written. Musical intervals are proportions, and

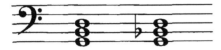

a. G major (*durus*), g minor (*mollis*)

b. C major (*durus*), c minor (*mollis*)

c. E♭ major (*mollis*), e minor (*durus*)   FIGURE 8.1. Durus and mollis triads

planetary motions form proportions in the sky. Kepler's use of ratios or proportions differs notationally from current practice; he usually separates the terms with a period rather than use a colon or write them as a fraction. I do not follow this convention. Other differences, less obvious but potentially more confusing, are these:

1. Kepler does not think of the terms of a proportion as ordered—after all, the order of two musical notes does not affect the harmony between them—but speaks of the single proportion between any two numbers. He almost invariably writes the smaller of the two first, thus 5:8 but not 8:5.

2. He uses addition and subtraction in combining proportions, where we would say that the operations involved are multiplication and division of fractions. It is, of course, entirely natural to think of the addition and subtraction of musical intervals, even though the arithmetic used to calculate the resulting interval involves multiplication or division.

3. He considers a "greater" proportion to be one that is farther from equality. The proportion 2:3 is thus greater than 3:4. Although this convention reverses one's intuitive sense of greater and lesser, it again corresponds to the ordering of musical intervals. A fifth (2:3) is a larger interval than a fourth (3:4).

For example, Kepler carries out the calculation $\frac{1}{2} \cdot \frac{2}{3} = \frac{1}{3}$ as 1:2 + 2:3 = 1:3. The inverse equation, $\frac{1}{3} \div \frac{2}{3} = \frac{1}{2}$, becomes 1:3 − 2:3 = 1:2.

Kepler's notation, compared to ours, is doubly inconvenient. First, it applies to proportions as a special class instead of subsuming them under the category of rational numbers. His is not a notation of fractional numbers but specifically a notation of proportions. Second, it excludes improper ratios, ratios greater than one. When subtracting ratios, he always subtracts the smaller ratio, the one closer to unity, from the larger ratio. This restricted arithmetic of proportions is reminiscent of ordinary arithmetic without negative numbers. The calculator must keep track of which terms are larger than others.

When dealing with the difference between proportions that are approximately equal—as he often is, necessarily, when comparing astronomical results with the harmonic proportions nearest to them—Kepler usually expresses the error as a ratio of adjacent integers, such as 62:63 or 36:37. Such proportions are called *epimoric* in Greek harmonic theory. They have special status in Ptolemy's *Harmonics*, and Kepler prefers them when representing "small" proportions, those close to unity. He does not ever explicitly discuss how large an error can be neglected in the comparison of harmonies to nature. In book five, chapter 4, he implicitly counts intervals as harmonic that deviate from a harmony by less than about 58:59. In his preliminary investigations into planetary scales in chapter 5 he recklessly threatens to ignore discrepancies less than a semitone, 15:16. In practice he seems to be comfortable disregarding errors smaller than about 50:51 or so. More than once he mentions the accuracy with which musical notes can be distinguished by the ear as the threshold of sufficiently good harmony.

Composition of the *Harmonice mundi*

Kepler's correspondence yields surprisingly little information on the composition of the *Harmonice mundi*. In 1599 Kepler had written to Herwart that he planned to compose a little work, *De harmonice mundi*, to consist of five "books or chapters," four of which had titles that correspond quite well to the eventual five books he published twenty years later. (Book 2 was to be about harmonic ratios or numbers but turned out to be about congruent figures in the book.) This project was much less ambitious than it became later, after Kepler's astronomical discoveries had provided a truly comprehensive and elegant basis for harmonic speculation. At any rate it remained incomplete. In January and again in November of 1607, while coaxing Ptolemy's *Harmonics* out of Herwart, Kepler reported that several years earlier he had actually begun to write the five books on harmonics but that he had wanted to finish his astronomical studies before completing the work.[6]

One simply does not know how much of this early work on harmonics was incorporated into the *Harmonice mundi* that Kepler published in 1619. The similar but not identical division of the subject into five parts indicates that he probably considered the *Harmonice mundi* as a perfected version of whatever he had written in Prague on harmonics. Book 2 was evidently reconceived altogether, judging from the change in title. Book 5, our present concern, must have changed radically, simply because Kepler learned an immense amount about the motions of the planets after writing his first draft on harmonics. The most important discovery in this regard, and one of the few events in the story that we can date precisely, was the "harmonic" law of planetary motion: the exact proportionality between the square of the period and the cube of the mean solar distance for all the planets. Kepler fully realized the importance of this result and carefully recorded the dates. On 8 March 1618 he first had the idea but rejected it because of a computational error. On 15 May he tried again and got it right.

One can scarcely imagine the difficulty of incorporating the period-distance relation and its consequences into an argument as tightly woven as that in the final chapters of the *Harmonice mundi*. Kepler noted at the end of that work that he had completed it on 27 May 1618, less than two weeks after his discovery of the relation, but that he continued to revise book 5 until 19 February 1619, while type was being set for that book.[7] It is probably safe to assume that most of the intervening months were devoted to rewriting chapter 9 in order to exploit the harmonic law.

PRESUPPOSITIONS

We are certainly unable, today, to recover the assurance with which Kepler sought the handwriting of the Creator among the gross phe-

[6] Kepler to Herwart, 14 December 1599, no. 148 in *G.W.* 14.100.11–20. Kepler to Herwart, January 1607, no. 409 in ibid., 15.389.105–12; and Kepler to Herwart, 24 November 1607, no. 461 in ibid., 16.79.57–67.

[7] Ibid., 6.368.24–26.

nomena of the natural world. It may indeed no longer be possible to combine a scientific mind of the highest order (and a determination to study the most profound questions about the natural world) with an untroubled belief in a world created for the sake of mankind. We have lost the ability to believe that divine purposes not only exist but are openly accessible to human reason. To understand Kepler's theories about the design of his world, we must be conscious of certain presuppositions that came naturally to him in the seventeenth century but that can be sustained today only with deliberate effort. I call these things presuppositions, not assumptions, because it is unlikely that Kepler deliberately postulated them. Probably he never doubted them.

The most important presuppositions were theological. Kepler shared with all European Christians of his age the belief that God, who had created the universe, had likewise created the human race in his own image. Mankind had a special place in the plan of Creation, and the Earth was likewise singled out, as the dwelling place of mankind, and the site of the Incarnation. For Kepler it was no longer located at the center of the universe, as it had been in ancient and medieval science and philosophy; yet it remained a very special place. If God had become man on Earth, then from Earth man should surely be able to detect the outlines of the great design of creation.

This fundamental belief implied, to Kepler and to others, that the universe had been constructed according to a plan that was intelligible, in principle, to human reason. Kepler differed from most of his contemporaries only in believing that he had actually discovered that plan.

I do not have much to say about these beliefs. They were not part of Kepler's contribution to the study of the natural world but rather part of the environment within which he carried out his studies. In fact, most of the great astronomers who followed Kepler in the seventeenth century (for example, Huygens, Halley, and Newton) shared his belief that where observations lacked, one could presume an orderliness in the dimensions of the created universe. Less systematic than Kepler in developing the implications of this belief, but supplied with telescopic observations, they were as matter-of-fact as he in supposing that the dimensions of the universe were determined according to a rational plan.[8]

---

[8] Some remarkable evidence for this assertion can be found in van Helden, *Measuring the Universe* (Chicago, 1985), chaps. 9–14.

These presuppositions are no longer part of our intellectual environment. We must provisionally reinstate them before attempting to understand Kepler, because apart from them the *Harmonice mundi* is baffling and absurd. To the twentieth-century mind it seems little more than a great pile of ingenuity heaped onto a mildly interesting substratum of coincidence. For Kepler, though, coincidence was not a part of the natural world. His world had been created by an omniscient and omnipotent God, and any pattern that could be detected in Creation was there because the Creator had put it there.

Book 5 of the *Harmonice mundi*

BOOK 5 of the *Harmonice mundi* opens with a *prooemium,* or preface, in which Kepler exults over his recent discovery of the relationship between the orbital radii of the planets and their periodic times: "At last . . . I brought it into the light, and beyond what I had ever been able to hope, I laid hold of Truth itself: I found among the motions of the heavens the whole nature of Harmony, as large as that is, with all of its parts as explained in book 3. It was not in the same way which I had expected—this is not the smallest part of my rejoicing—but in another way, very different and yet at the same time very excellent and perfect."[1]

During the years when the "restitution of the motions" (the calculation of orbital parameters for all the planets, in conformity with his New Astronomy) forced Kepler to suspend his harmonic speculations, he received and read Ptolemy's *Harmonics,* as discussed in chapter 7. He was astounded at the agreement between Ptolemy's fifteen-hundred-year-old contemplations of celestial harmony and his own but recognized that no one in the ancient world could possibly have grasped the full truth in these matters.

> To be sure, much was still lacking in the astronomy of that age; and Ptolemy, having begun badly, could plead desperation. Like the Scipio of Cicero, he seems to have recited a kind of Pythagorean dream rather than advancing philosophy. Both the primitive state of ancient astronomy and the wonderfully exact agreement between the two meditations, at an interval of fifteen centuries, confirmed my intention of pressing onward. What more is there to say? The very nature of things was revealed to men through the different interpreters of distant ages; it was the finger of God, as the Hebrews say, which conceived the same thing here in the minds of two men who gave themselves totally to the contemplation of Nature, concerning the structure of the world; since neither of the two led the other into this path.[2]

[1] *G.W.* 6.289.13–19.
[2] Ibid., 6.289.27–38.

Kepler is eager to assure the reader that he sought music in the heavens on his own inspiration, and not as a mere commentator; and one cannot doubt him. He continues with an exultant passage that reveals his joy at having discovered the harmonic structure of the heavens.

> Now—eighteen months after the first light, three months after the daybreak, and very few days indeed after the sunrise of a most admirable contemplation—nothing holds me back, I can indulge the sacred passion, I can scoff at mortals with the frank confession that I am stealing the golden vessels of the Egyptians, to build with them a tabernacle for my God far from the bounds of Egypt. If you indulge me, I will rejoice; if you are angry, I will bear it. I throw the die and write the book; it matters not whether it will be read now or in the future. Let it wait a hundred years for its reader, as God himself has awaited a witness for six thousand years.[3]

Caspar takes the sunrise described here as a reference to the discovery of the Third Law, and that seems likely, given the precision with which Kepler describes the event. The *prooemium* was not written after the extensive revisions which that discovery provoked in book 5; because it was printed with the initial pages. (It shares page 179, in the 1619 pagination, with the table of contents for book 5, which was written when Kepler still expected to print Ptolemy's *Harmonics* as an appendix. That table of contents shares page 180 with the beginning of chapter 1. There is no break in the pagination where the *prooemium* might have been inserted, if it had been printed last.) Kepler tells us that he discovered the period-distance relation on 15 May 1619, finished the *Harmonice mundi* on 27 May 1618, and revised book 5 over the next nine months while type was being set. The *prooemium* was surely written during the few days between the fifteenth and the twenty-seventh, before the initial euphoria gave way to the grinding labors of rewriting chapter 9.

Book 5 is entitled "Concerning the Most Perfect Harmony of the Celestial Motions." It describes the large-scale structure of the physical world and the harmonies that guided the creation of that structure. For Kepler the physical world consists of an actual sphere of the fixed stars; the Sun, at its center; and the six planets revolving around the Sun in the intermediate region. Ratios of distances among the Sun, the

---

[3] Ibid., 6.289.38–290.9.

planetary orbits, and the sphere of the stars are thus the most notable proportions in the world. Because the sphere of the stars is so distant, and because one observes it from very near its center, any hypothesis about its relative size is necessarily speculative. (That it is very, very large is an essential postulate of Copernican cosmology.) Kepler believes, as he wrote elsewhere, that the sphere of Saturn might well be a mean proportional between the solar body and the sphere of the fixed stars, thus putting the three components Sun, planetary system, and stars into a proportion that would surely please the Creator.[4] In this book he deals with matters of more certain knowledge: proportions within the solar system, which can actually be observed, or rather deduced from astronomical observation.

CHAPTER 1: REGULAR POLYHEDRA

Proportions within the planetary system itself are more numerous, and easier to estimate precisely, than those involving the sphere of the fixed stars. Kepler's first hope, in the *Mysterium cosmographicum*, was to explain these proportions by means of the regular polyhedra. He had since realized that this theory was both inaccurate and inadequate. Each planet varies its distance from the Sun, and among the greatest, least, and perhaps mean distances of six planets there are far more proportions than the polyhedral hypothesis accounts for. By the time he is writing the *Harmonice mundi* Kepler confidently believes that geometrical harmonies—harmonic ratios that give rise to musical harmonies—provide the underlying theory, the *ratio*, for all of these proportions. The five regular polyhedra are still there, providing approximate proportions among the planetary spheres. What is more important, they determine the exact number of planetary spheres: six. Kepler remains firm in his belief that the regular polyhedra explain the number and approximate sizes of the planetary orbits. He therefore opens book 5 with two chapters summarizing the pertinent characteristics of the regular polyhedra.

He classifies the regular polyhedra into primary and secondary figures. The primary figures are those with solid angles formed from three lines, and they are the cube, the tetrahedron, and the do-

---

[4] For example, *De stella nova* (1606), G. W. 1.234–35; and *Epitome astronomiae Copernicanae*, book 4 (1620), ibid., 7.286. A. van Helden's *Measuring the Universe* (Chicago, 1985) is now the standard account of attempts to estimate the proportions of the universe.

decahedron. Of these the cube takes primacy because it is "firstborn, and having the character of a whole, by the very form of its generation." In other words, as he explained in the *Mysterium cosmographicum* and the *Epitome*,[5] the cube can be generated from one of its sides and a right angle. The tetrahedron is next in dignity, since it can easily be cut out of a cube, while the dodecahedron is last, as it must be constructed from pieces that are cut out of a cube. (Kepler's figures, reproduced as figure 9.1, show how to cut up and reconstruct the regular polyhedra better than any explanation.) And that was the order in which the primary figures have been placed among the planetary spheres, starting from the outside. The cube came between Saturn and Jupiter, the tetrahedron came between Jupiter and Mars, and the dodecahedron came between Mars and Earth.

The secondary figures are those that have solid angles formed from more than three lines, and they are the octahedron and the icosahedron. The octahedron is nobler, since it is closely related to the cube, and hence it comes first among the planetary spheres starting from the inside, between Mercury and Venus. The icosahedron is left with the interval between Venus and Mercury.

Kepler associates two of the primary figures, as males, with the two secondary figures, as females, in "marriages" (*conjugia*). The relationship is literally one of inscription, "as women will be assigned to, and in a way subjected to [*inscribuntur et veluti subjiciuntur*] men." An octahedron is inscribed in a cube by connecting the centers of the six faces of the cube; these two figures thus form the "cubic marriage." An icosahedron is inscribed in a dodecahedron by connecting the centers of the twelve faces of the latter; these figures form the "dodecahedric marriage." The two figures in each marriage have equal ratios of circumscribed sphere to inscribed sphere. The tetrahedron is left outside the marriages, but one tetrahedron can be inscribed inside another tetrahedron, so Kepler calls it "unmarried or androgynous."[6]

CHAPTER 2: HARMONIC PROPORTIONS

In the second chapter of book 5 Kepler enumerates all the proportions (and there are a lot of them) that can be formed systematically from

---

[5] Chap. 5 of the *Mysterium* (*G.W.* 1.31–32 or 8.53–54); book 4 of the *Epitome* (ibid., 7.268).

[6] Kepler's imagery is weakened by the reciprocity of these inscriptions: the primary figures can equally well be inscribed in the secondary figures, as he knew. *G.W.* 7.271.13–18; Field, *Kepler's Geometric Cosmology* (Chicago, 1988), p. 54.

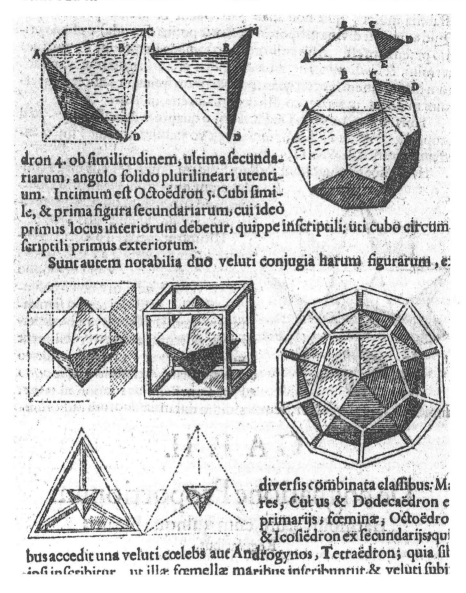

dron 4. ob similitudinem, ultima secundariarum, angulo solido plurilineari utenti-
um. Intimum est Octoëdron 5. Cubi simi-
le, & prima figura secundariarum, cui ideò
primus locus interiorum debetur, quippe inscriptili; uti cubo circum
scriptili primus exteriorum.

Sunt autem notabilia duo veluti conjugia harum figurarum, e:

diversis combinata classibus: M:
res, Cubus & Dodecaëdron e
primarijs; foeminæ, Octoëdro
& Icosiëdron ex secundarijs;qui

bus accedit una veluti coelebs aut Androgynos, Tetraëdron; quia si
insi inscribitur ut illæ foemellæ maribus inscribuntur.& veluti subi

FIGURE 9.1. Construction of the regular polyhedra

the regular polyhedra. Each figure is well provided with ratios: ratios
among the numbers of edges, faces, plane angles, and solid angles;
ratios among the numbers of edges per face and per solid angle; ratios
between the edge-length and the radius of the circumscribing sphere;
and so on, in great profusion. The most important of these ratios is the
"proportion of the spheres," the ratio between the radius of the in-

scribed sphere and that of the circumscribed sphere. This ratio is 1:3 for the tetrahedron, 1:$\sqrt{3}$ for the figures in the cubic marriage, and "ineffable"—a little greater than 4:5, actually 1:$\sqrt{15 - 6\sqrt{5}}$—for the figures in the dodecahedric marriage.

Next Kepler decides, on abstract grounds, which of the regular polyhedra is most suited to each of the harmonic proportions. With each polyhedron carrying such a heavy load of proportions, he concentrates on the proportions of the spheres, assigning each of the important harmonic proportions to the polyhedral marriage whose proportion of spheres is closest to that harmonic proportion. The proportion 1:2 is a little greater, and the proportion 3:5 a little less, than 1:$\sqrt{3}$, so these proportions are assigned to the cube and octahedron. The harmonic proportions 3:4 and 5:8 are a little greater, and the proportions 4:5 and 5:6 a little less, than the proportion of spheres in the dodecahedric marriage, so those proportions are assigned to the dodecahedron and icosahedron.

Assignment of harmonic proportions to the tetrahedron is not so easy. Its proportion of spheres is 1:3, which is perfectly harmonic, an octave plus a fifth; but Kepler hesitates. If "for certain reasons" the harmonic proportions 1:2 and 1:3 belong to the cube, then the tetrahedron should get the doubles (that is, the squares) of these proportions, 1:4 and 1:9, because the tetrahedron's proportion of spheres is double the cube's proportion of spheres.[7] Setting aside for the moment the reasons—unstated in chapter 2—for assigning 1:3 to the cube, instead of to the tetrahedron, whose proportion of spheres is exactly 1:3, one still faces a novel way of reasoning. Instead of looking directly at the tetrahedron's proportion of spheres, Kepler fixes his attention on that proportion being "double" the cube's proportion of spheres. Even so, he must adjust the results, because 1:9 is not a harmonic proportion. The nearby 1:8, he decides, can step in alongside 1:4 as a suitable harmony for the tetrahedron.[8]

As is often the case, the reader must indulge Kepler in these "argu-

---

[7] *G.W.* 6.296.5–9. The reasons turn up in prop. 8 of chap. 9.

[8] Kepler tries to apply this second way of reasoning to the dodecahedral marriage also: its common proportion of spheres is, he says, a third that of the cubic marriage. Hence it might be that 4:5 and 3:4, harmonic proportions about a third of the cube's 1:2 and 1:3, should be assigned to the dodecahedron and the icosahedron. Ibid., 6.296.10–15. There appears to be no trace of this possibility in later chapters of the *Harmonice mundi*. (Three times the proportion 1:$\sqrt{15 - 6\sqrt{5}}$ is actually quite close to 1:2, but not very close to 1:$\sqrt{3}$, the *orbium proportio* of the cubic marriage.)

ments." Even if one wants to be convinced, one cannot be convinced by the *Harmonice mundi*. The greatest impediment to understanding this book is one's natural expectation that Kepler is trying to convince his readers that the astronomical data imply a harmonic design in the heavens. Expecting this, one is sure to be disappointed. Kepler's intentions are more modest. He is manufacturing logic that would lead to results already known. *Of course* he is justifying known results; he can scarcely pretend to have deduced the proportions of the heavens apart from astronomical knowledge. In large part, the purpose of book 5 is to show that a harmonic logic exists in the known facts of astronomy. He is not proposing a design for an unknown planetary system but rather finding the design of the one in which he lives and showing that it is consistent with the highest wisdom.

Along with the five regular polyhedra known from Euclid, Kepler introduces a sixth, which he is generally credited with discovering. This is a solid star figure or "hedgehog" (*echinus*), shown in figure 9.2, which is formed by extending the twelve pentagonal sides of the dodecahedron into five-pointed stars. The points of the stars fit together perfectly to form five-sided pyramids resting on each of the faces of the original dodecahedron. If one regards the five-pointed star as a

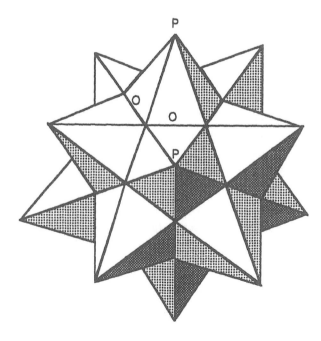

FIGURE 9.2. A regular hedgehog

regular five-sided polygon with interior angles of 36° (and intersecting sides), then the *echinus* is a polyhedron with twelve identical regular polygonal faces and twelve identical solid angles: hence a new regular polyhedron, one unknown to the ancients. It is by no means as simple a polyhedron as the other five, because of all the intersections and because neither it nor its star-shaped face is convex; but Kepler is justly pleased with it and tries to work it into his theory where he can. It fits best between Mars and Venus.

CHAPTER 3: ASTRONOMICAL THEORY

In the third chapter of book 5 Kepler provides "a summary of the astronomical theory necessary for the contemplation of the celestial harmonies." He bluntly rejects the Ptolemaic hypotheses as false and urges his readers to accept the Copernican hypotheses. The coming harmonic speculations could not be derived in a geocentric universe. The astronomical data on which they rest, the relative distances of the planets from the Sun, are left undetermined in Ptolemy's system. Copernicus, by recognizing that the apparent motions of all the planets share a component that is due to the Earth's motion around the Sun, was able to measure the orbits by comparing the size of this common component to the size of the components proper to each planet. In doing so he unified all the planetary models into a solar system whose parts and motions have ascertainable proportions among them. These proportions are the basis for all Kepler's speculations in book 5, which consequently would be utterly without foundation to a geocentric astronomer.

He lists the requisite astronomical results in thirteen propositions. They are essential to understanding the harmonic investigations that followed, so we paraphrase them here:[9]

> 1. All the planets move around the Sun, except the Moon, which moves around the Earth. To the traditional planets Kepler appends the Earth, which adds a sixth circle around the Sun—either actually, by moving around the Sun, or implicitly, "itself unmoving, and the whole planetary system rotating." (He has dismissed the adherents of Ptolemy but is unwilling to lose those of Tycho.)

[9] The following is from Kepler's summary of astronomical theory in chap. 3, *G.W.* 6.297–305.

2. The orbits are eccentric to the Sun, so that each planet has a greatest distance and a least distance, and passes through all the intermediate distances.

3. The number of planets, six, is determined by the five regular polyhedra, as Kepler first showed in his *Mysterium cosmographicum*.

4. The proportions of the orbits are approximately determined by the interposition of the polyhedra between them, in the order given above. The inexactness is not due to the Creator's sloppiness. Indeed, the polyhedra alone could not possibly have determined the distances of the planets from the Sun, since the orbits are eccentric. Between two planets, four comparisons of distances are possible, considering only the greatest and least distance of each from the Sun. A single polyhedron provides only a single ratio. Kepler concludes that "other principles, beyond the regular polyhedra, are needed to establish the orbits and the diameters and the eccentricities all together."

The Motion of a Planet

About the true motion of a planet, Kepler knows a great deal. In propositions 5–13 of chapter 3, he frankly presents his own doctrine, for there is no alternative. He needs to talk about the speed of the planets' motions, and his is the only theory to give a coherent account of the actual speed of planetary motion.

Kepler's theory, as he summarizes it here, is built on geometry rather than calculus and therefore lacks any formal concept of instantaneous velocity. Proposition 5 condenses into one paragraph the core of his planetary theory. He begins by stating, in his own preferred form, the relation we know as his area law:

5. A planet's unequal delays in equal parts of the eccentric follow the proportion of its distances from the sun, the source of motion.

The delays (*morae*) of which Kepler speaks here are simply the amounts of time required to traverse the parts of the eccentric. He is thus asserting that the time required for the planet to move through a section of "the eccentric," of specified size, is proportional to its distance from the Sun. He also states the relation in inverse form, that "in equal times, . . . the true diurnal arcs of a single eccentric orbit" are in inverse proportion to their intervals from the Sun.

These relations can be read in two ways, and I suspect that Kepler wrote them so that they could be read in two ways. The straightforward reading is that the planet's velocity is inversely proportional to its distance from the Sun. This reading is clear, plausible, and almost true. During the early years of his work on Mars Kepler thought it true; but in his final conquest of Mars, as narrated in chapter 58 of the *Astronomia nova*, he did not so much revise his theory as redefine the terms of that theory. The newer, more precise reading is hidden within his words, to be understood by those of his readers who have grasped his planetary theory in all its detail. It is exactly true, but by no means as clear (or as plausible) as the straightforward reading. It depends upon the distinctions shown in figure 9.3.

The elliptical orbit is inscribed within an eccentric circle. The distance between the two is greatly exaggerated in the figure: in reality, for an orbit of large eccentricity such as that of Mars, whose eccentricity is almost a tenth of the orbital radius, the maximum width of the crescent-shaped "lunula" between ellipse and circle is only about 0.005 of the radius. Arc DE on the eccentric circle is projected perpendicularly toward the apsidal line FG onto arc AB of the ellipse. Kepler discovered while writing the *Astronomia nova*, and would demonstrate in book 5 of the *Epitome*, that for very small equal arcs such as DE on the eccentric, the planet's delay in the corresponding *unequal* arcs AB on the ellipse is exactly proportional to its distance SA from the Sun.[10]

To be precise, then, Kepler's statement that a planet's "delays in equal parts of the eccentric observe the proportion of its intervals from the Sun" refers not to parts of the orbit but to parts of the eccentric circle, arcs through which the planet never passes! The planet's delays in those parts of the eccentric are the amounts of time it takes to pass through the corresponding parts of the orbit, where the correspondence is given by perpendiculars dropped to the apsidal line, as in figure 9.3. Kepler's demonstration, in the *Epitome*, that such delays are proportional to the planet's distance from the Sun is both precise and subtle; it is evidently too subtle for Kepler to explain in the *Harmonice mundi*. He therefore formulates his distance law in terms of the ambiguous word *eccentric*—which he himself often uses for the actual

---

[10] This is the final form of Kepler's "distance law," more commonly known today as the area law. The point of taking equal arcs on the eccentric circle is that the corresponding unequal arcs on the ellipse turn out to have equal components around the Sun, if one neglects the radial motion. He demonstrated this in book 5 of the *Epitome*, not published until 1621. See Stephenson, *Kepler's Physical Astronomy* (New York, 1987), pp. 161–65.

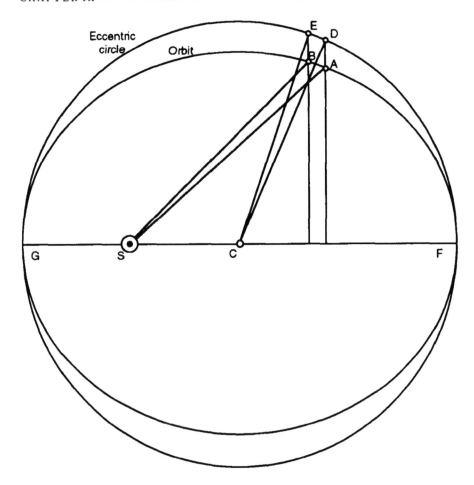

FIGURE 9.3. Arcs of the orbit and the eccentric

orbit—and leaves it to his readers to distinguish orbital arcs such as AB from eccentric arcs such as DE if they know how.

In this same meaty paragraph, Kepler continues proposition 5 by stating flatly that "it has been demonstrated by me that the orbit of a planet is elliptical, and that the Sun, the source of motion, is in one of the foci of this ellipse." The planet, he goes on, is precisely at mean distance from the Sun when it has traveled one-quarter of the whole circuit starting from aphelion. At the quadrants of the orbit, then, the "daily mean motion of the planet in the eccentric is the same as its true diurnal arc of the eccentric." (Once again, neither term of this equation refers to motion on the actual orbit.) He concludes the paragraph with several further *axiomata* about the true motions of planets, which follow immediately from the distance law on an elliptical orbit.

After the compact summary of a planet's true motions in proposition 5, Kepler turns to the apparent motions.

> 6. Among equal true motions, that which is farthest from the center of the world, would appear smallest if seen from that center [the Sun]. Thus aphelial motions would appear smaller than perihelial motions, even if the true motions were equal. Since the true motion at aphelion is less than that at perihelion, in proportion to the distance, it follows that the proportion of the apparent sizes of two arcs from a single eccentric is very nearly the inverse square of the proportion of their distances from the Sun.

Kepler himself points out that this conclusion is approximate, and qualifies the approximate inverse-square proportionality in proposition 6 with two explicit stipulations:[11] first, that the arcs of the eccentric are not large; and second, that the planet's eccentricity is not very great. The first stipulation is purely technical: he needs small arcs because, lacking calculus, he conceives and expresses his theory in terms of arcs of finite size. The second stipulation is somewhat more interesting, and for it he gives two reasons, one mathematical and the other physical. Strictly speaking, increased distance produces a proportionate optical reduction in the sine of an apparent angle rather than in the angle itself. This consideration, Kepler correctly claims, is of no consequence for small arcs at great distance. Of somewhat more importance is the reduction in the apparent angle when an arc is seen obliquely—which is true around the quadrants, and particularly so when the orbit is greatly eccentric.[12]

> 7. Motions seen from the Earth, rather than from the Sun, play no role in the celestial harmonies. They cannot; for the Earth is not the source of the planets' motion. Furthermore, the illusory retrograde motions observed from the Earth would throw all the proportions into confusion.

[11] We may note that here, also, Kepler silently passes over the distinction between eccentric and orbit. The distances that govern the physical shortening of the diurnal arcs on the eccentric are distances to the planet itself on its elliptical orbit; but the distances that govern the apparent shortening of eccentric arcs seen from a greater distance are distances to those arcs themselves, on the eccentric.

[12] *G.W.* 6.300.31–301.22. Later, in chap. 4, Kepler alludes to this qualification as a reason why the extreme apparent motions of Mercury do not correspond exactly to an inverse-square proportionality applied to its mean apparent motion. Ibid., 6.313.1–12.

Since the geocentric planetary motions "degenerate" to zero, and then reverse themselves, all possible proportions would equally be attributable to them. Certainty can be sought in the proportions only if one disregards the illusions arising from the Earth's motion and uses either the Copernican or the Tychonic system.

Motions of More Than One Planet

The propositions to this point have dealt with individual planets. To consider relations between different planets, Kepler defines some terms. The "near apsides" of two planets are the perihelion of the superior planet and the aphelion of the inferior (regardless of whether these two points are actually near each other in the zodiac). "Extreme motions" are the slowest and fastest motions of a planet. The "converging or converse" extreme motions of a pair of planets are those in the near apsides, namely the perihelial motion of the superior planet and the aphelial motion of the inferior planet. The "diverging or diverse" extreme motions are the other two, the aphelial motion of the superior planet and the perihelial motion of the inferior. The proportion of converging motions of two planets is therefore the smallest proportion that could occur between their motions, and the proportion of diverging motions is the largest. These two proportions are singled out among the proportions possible between the motions of two planets, and hence one can expect them to have been given special attention by the Creator.

At this point, in chapter 3 of book 5, Kepler inserts his first statement of what we know today as his third or harmonic law of planetary motion.[13] He appends a curious qualification to it:

> 8. It is most certain and exact, that the proportion between the periodic times of any two planets is precisely the sesquialterate of the proportion of the mean distances of the orbits themselves; with this caution, that the arithmetic mean between the two diameters of the elliptical orbit must be a little less than the longer diameter.

[13] Ibid., 6.302.14–24. Kepler did not discover this relation until the *Harmonice mundi* was nearly complete, on 15 May 1618. He writes that he finished the whole book on 27 May and finished revising book 5, presumably to take account of his new discovery, while the type was being set, on 19 February 1619: ibid., 6.368.24–26.

The proportion of periodic times is one and one-half times the proportion of the mean diameters; in modern terminology, the proportion of times equals the proportion of diameters raised to the $\frac{3}{2}$ power.

The qualification about the arithmetic mean of the diameters is hard to understand. That the "mean between the two diameters . . . must be little less than the longer diameter" is a roundabout way of saying that the two axes of the ellipse must be nearly equal—in other words, that the eccentricity must be small. In a side note, Kepler gives a clue to his intentions, repeating from the *Astronomia nova* his contention that the mean of the two diameters of an ellipse either equals or is slightly less than the diameter of a circle with the same circumference as the ellipse.[14] This suggests that Kepler's qualification of his harmonic law is intended to ensure that the elliptical circumference of the orbit is not very much shorter than the circumference of the eccentric circle from figure 9.3. This probably reflects an early stage in his attempts (attempts that I have not been able to reconstruct) to understand the harmonic law.

Probably Kepler in 1618, trying furiously to assimilate the new discovery into his harmonic speculations and physical theories, was not altogether certain whether the harmonic law was itself one of the Creator's archetypal principles (and hence exact "by construction") or whether it was instead a consequence of other archetypes. In the latter case, the simplicity of the archetypal proportions could have been compromised by other causes along the way. Later in book 5, when comparing the results of the harmonic law with orbital parameters calculated from observations, Kepler found that "in Mercury alone there is a slight discrepancy."[15] Now, Mercury has an eccentricity more than double that of any other planet, and Kepler was quite conscious, from his labors in the *Astronomia nova* years, that simple relations could turn out to be approximations, valid only for orbits of small eccentricity. The qualification he places upon his first statement of the harmonic law perhaps reveals his uncertainty whether it, too, is a small-eccentricity approximation.

Having cautiously inserted one of his most profound discoveries,

---

[14] The reference to the *Astronomia nova* is to *G.W.* 3.306.36–307.24. Riccioli quoted this statement of the harmonic law in full, including the qualification about the arithmetic mean of the two diameters, in his *Almagestum novum*, vol. 2, p. 532.

[15] *G.W.* 6.359.3–4.

Kepler continues with several corollaries applying it to the true and apparent extreme diurnal motions of the planets. In these propositions he spells out what is implicit in the period-distance relation and the distance law: that the proportion of the converging or diverging motions of a pair of planets, together with the proportion of their periodic times, determines the extreme distances corresponding to those motions, and hence determines the eccentricities. Some of the propositions are clumsily stated and show signs of being thrown together hastily, as one might expect under the circumstances.

9. To measure the true diurnal motion of a planet, multiply its true [angular] diurnal arc by its mean distance from the Sun.

10. The apparent diurnal motions are proportional to the true diurnal motions divided by the corresponding distances from the Sun.

11. The apsidal distance of a planet equals its mean distance multiplied by the ratio of the planet's mean apparent motion to the mean proportional between that mean motion and the apsidal motion.

If we use M and R to represent a planet's apparent motion and distance at either apse, and $\bar{M}$ and $\bar{R}$ to represent its mean apparent motion and mean distance, then this eleventh proposition is that:

$$\frac{R}{\bar{R}} = \frac{\bar{M}}{\sqrt{M \cdot \bar{M}}}$$

$$\text{Hence } \frac{R^2}{\bar{R}^2} = \frac{\bar{M}}{M}$$

which simply means that the apparent motions at the apsides are inversely as the square of the distance from the Sun, as has already been stated in proposition 6. The ratio of apsidal distance to mean distance, of course, determines the planet's eccentricity, so this proposition allows Kepler to calculate the eccentricity from a planet's mean and extreme motions.

12. The mean apparent motion of a planet is less than the geometric mean of the planet's two extreme apparent motions, by half as much as that geometric mean is less than the arithmetic mean of the two extreme motions.

Thus, Kepler continued, if a planet's aphelial and perihelial motions are as 8 to 10, the geometric mean between them would be $\sqrt{80}$, the

BOOK FIVE OF THE *HARMONICE MUNDI*

arithmetic mean would be 9, and the planet's mean motion would be $\sqrt{80} - \frac{1}{2}(9 - \sqrt{80})$. Caspar has shown[16] that proposition 12 can be derived as an approximate consequence of proposition 6, on which its predecessor also depends. I suppose that Kepler came upon this hardly obvious result empirically, by making trial calculations.

The thirteenth and final proposition of chapter 3, relating the apparent extreme motions of a pair of planets, is still more abstruse:

> 13. The proportion of the superior planet's apparent perihelial motion to the inferior planet's apparent aphelial motion is less [closer to equality] than the inverse $\frac{3}{2}$ power of the proportion of the corresponding distances. Moreover, as long as the combination of the ratios of extreme distance to mean distance for the two planets is less than the square root of their ratio of mean distances, the proportion of their extreme motions is greater than the proportion of their extreme distances. On the other hand, if the former ratio is greater than the square root of the ratio of mean distances, the proportion of extreme motions would be less than the proportion of extreme distances.

To follow Kepler's demonstration of this we adopt the following notation. The superior planet's mean and perihelial distances from the sun are R and R_p; the inferior planet's mean and aphelial distances are r and r_a; the superior planet's mean and perihelial motions (apparent velocities) are M and M_p; and the inferior planet's mean and aphelial motions are m and m_a. We retain the additive language for combining proportions, placing the smaller term of a proportion first as Kepler did. These conventions allow us to follow Kepler's demonstration directly but without constant reference to his unhelpful diagram.[17]

The first part of proposition 13 is that

$$M_p{:}m_a < \tfrac{3}{2}r_a{:}R_p$$

Kepler did not bother to demonstrate this. It is true because

$$M{:}m = M{:}M_p + M_p{:}m_a + m_a{:}m$$
$$M_p{:}m_a = M{:}m - (M{:}M_p + m_a{:}m) \tag{1}$$

By the sixth proposition of this chapter, the proportion between a planet's apparent motions at two points in its orbit is twice the inverse proportion of its distances from the Sun. At mean and converging distances for each of the planets one has the following two relations:

[16] Ibid., 6.548–49.
[17] Ibid., 6.304.6–305.7. Caspar's notation is given in his notes: ibid., 6.547.

$$M{:}M_p = 2R_p{:}R$$
$$m_a{:}m = 2r{:}r_a$$

Adding, $M{:}M_p + m_a{:}m = 2(R_p{:}R + r{:}r_a)$ (2)

Substituting this into the right side of (1):

$$M_p{:}m_a = M{:}m - 2(R_p{:}R + r{:}r_a)$$
$$= M{:}m - 2(r{:}R - r_a{:}R_p)$$

The commonplace operation with proportions executed here amounts to the replacement of $\frac{w}{x} \cdot \frac{y}{z}$ with $\frac{y}{x} \div \frac{z}{w}$.

By the period-distance relation we have:

$$M{:}m = \tfrac{3}{2}r{:}R \tag{3}$$

So $M_p{:}M_a = \tfrac{3}{2}r{:}R - 2r{:}R + 2r_a{:}R_p$
$$= -\tfrac{1}{2}r{:}R + 2r_a{:}R_p \tag{4}$$

Obviously the ratio of mean distances is greater (in Kepler's sense, meaning more distant from equality) than the ratio of converging distances; that is, $r{:}R > r_a{:}R_p$. So

$$M_p{:}m_a < \tfrac{3}{2}r_a{:}R_p \hfill Q.E.D.$$

The second part of the proposition is conditional upon an assumption about the eccentricities. Assume that:

$$R_p{:}R + r{:}r_a < \tfrac{1}{2}r{:}R \tag{5}$$

Then for some certain difference (*defectus certus*) d it will be true that:

$$R_p{:}R + r{:}r_a = \tfrac{1}{2}r{:}R - d \tag{6}$$

Substituting this expression into (2):

$$M{:}M_p + m_a{:}m = r{:}R - 2d$$

Subtracting this equation from the period-distance relation (3):

$$M{:}m - M{:}M_p - m_a{:}m = \tfrac{3}{2}r{:}R - r{:}R + 2d$$
$$= \tfrac{1}{2}r{:}R + 2d$$

But since $M{:}m = M{:}M_p + M_p{:}m_a + m_a{:}m$, one can reduce the left side to obtain simply:

$$M_p{:}m_a = \tfrac{1}{2}r{:}R + 2d \tag{7}$$

But $r{:}R$ can also be divided into its constituent parts:

$$r{:}R = r{:}r_a + r_a{:}R_p + R_p{:}R$$
$$= (R_p{:}R + r{:}r_a) + r_a{:}R_p$$

Substituting (6) into this gives:

$$r{:}R = \tfrac{1}{2}r{:}R - d + r_a{:}R_p \text{ or, equivalently,}$$
$$\tfrac{1}{2}r{:}R + d = r_a{:}R_p \tag{8}$$

Comparing this with (7), one sees that:

$$M_p{:}m_a = r_a{:}R_p + d \tag{9}$$
$$M_p{:}m_a > r_a{:}R_p \hfill Q.E.D.$$

Kepler remarks here that the inverse proposition, namely that $R_p:R + r:r_a > \frac{1}{2}r:R$ implies $M_p:m_a < r_a:R_p$, is proved using the same demonstration, by interchanging "greater" with "less," "plus" with "minus," and "excess" with "defect." We will forego the pleasure of confirming this.

Before closing chapter 3, Kepler states two corollaries for future use:

> 13.1. The proportion of two converging intervals is the mean of half the proportion of the spheres and the inverse proportion of the converging motions.

This complex-sounding statement is nothing but a rearrangement of equation (4).

> 13.2. The proportion of the diverging motions is much greater than the sesquialterate of the proportion of the spheres.

For the proportion of diverging motions is greater than the proportion of the mean motions, which itself equals the sesquialterate of the proportion of the spheres.

CHAPTER 4: THE PROPER AND EXTREME HARMONIES

Kepler begins his discussion of the harmonies in planetary motions by explaining how he had realized where such harmonies should be sought. Several numbers are eligible, it seems: a planet's distance from the Sun; its periodic time; its daily eccentric arcs; its daily delays (*morae*) in the eccentric arcs;[18] its apparent daily arc, or angular motion, seen as if by an observer in the Sun; and finally, its true daily path, measured along the path, rather than an arc of a circle. Kepler begins with the periodic times and has little difficulty showing that the Creator did not introduce harmonic proportions directly into the periods of the planets. The numbers simply are not in harmonic ratios.

He turns his attention to the delays. The periodic times are composed out of the planets' various delays in the different portions of

[18] Evidently a mistake. Daily delays in the eccentric arcs (*Morae in iis arcubus diurnae*, where *iis arcubus* refers to *Arcus eccentrici diurni*: G.W. 6.306.8) equal one day, by definition. From Kepler's later discussion (ibid., 6.308.23–29) it seems clear that he intended to say here "delays in equal arcs." Field (*Kepler's Geometric Cosmology* [Chicago, 1988], p. 146) has "length of daily path along these arcs," which comes to the same thing.

their orbits, so Kepler reasons that the agreeable geometric proportions—which must surely reside somewhere in the planetary system—are to be sought either in these delays or in something "prior to them in the mind of the Craftsman." Three possibilities, phrased differently, are in fact mathematically equivalent. For any individual planet, the *delays in equal eccentric arcs* are inversely proportional to the *diurnal arcs* (measured in degrees) and directly proportional to the *distances from the Sun*, so these three measurements would yield the same proportions. They all vary within a planet's orbit, but clearly the harmonies would have been established at the extremes: at aphelion and perihelion. These are the best-marked places in the orbit, as well as the places where delays, diurnal arcs, and distances changed most slowly.

Kepler accordingly lays out a table (table 9.1) with the aphelial and perihelial distances of the planets, here denoted *A* and *P* respectively, to check in one effort whether harmonies are to be found in the distances, the delays, or the diurnal arcs. The "proper proportions"

TABLE 9.1.
Solar Intervals Compared to Harmonies

1 Diverging proportion	2 Converging proportion	3 Interval		4 Proper proportion
		Saturn A	10,052	More than a minor tone,
		Saturn P	8,968	less than a major tone
$\frac{2}{1}$	$\frac{5}{3}$			
		Jupiter A	5,451	Not harmonic but about
		Jupiter P	4,949	$\frac{11}{10}$, or half of $\frac{6}{5}$
$\frac{4}{1}$	$\frac{3}{1}$			
		Mars A	1,665	$\frac{1,665}{1,388}$ would be $\frac{6}{5}$;
		Mars P	1,382	$\frac{1,665}{1,332}$ would be $\frac{5}{4}$
$\frac{5}{3}$	$\frac{27}{20}$			
		Earth A	1,018	$\frac{1,020}{980}$ is a diesis, so
		Earth P	982	this is less than a diesis
$\frac{10,000}{7,071}$	$\frac{27}{20}$			
		Venus A	729	Less than $1\frac{1}{2}$ comma,
		Venus P	719	more than $\frac{1}{3}$ diesis
$\frac{12}{5}$	$\frac{243}{160}$			
		Mercury A	470	More than $\frac{243}{160}$;
		Mercury P	307	less than $\frac{8}{5}$

between the extreme distances of individual planets (column 4) fail to reveal harmony. The diverging and converging distances of adjacent planets, in columns 1 and 2, show promise, however. Six out of ten proportions are close to harmonic. Saturn and Jupiter form an octave and a major sixth. Jupiter and Mars form a double octave and an octave-plus-fifth. Mars and Earth form a major sixth, although their converging interval is not harmonic. Earth and Venus fail to form any pleasing harmony; the best Kepler can say is that their diverging interval is about a half-octave, far from any harmony. Venus and Mercury, finally, form an octave-plus-minor-third as their diverging interval.

Kepler provides no explicit criterion, of course, for deciding when an observed proportion is "close enough" to a harmony. Six of the proportions he has entered in columns 1 and 2 of the table are actual harmonies. Among those six the greatest discrepancy with the distances in column 3 occurs for the diverging proportion of Mars and Earth, whose major sixth differs from the intervals in column 3 by about 58:59.

Curiously enough, three of the remaining four proportions in columns 1 and 2 are within reach of a harmony, although Kepler fails to claim them. He labels the converging proportions both of Mars-Earth and of Earth-Venus as 20:27, a very accurate characterization; but both are also close to the harmony 3:4, a perfect fourth. The Mars-Earth proportion errs by 55:56, so it is slightly worse than any of the harmonies that Kepler lists in this table, while the Earth-Venus proportion errs by a mere 97:98, and is hence more accurately harmonic than most of the proportions Kepler lists. Similarly, the converging proportion of Venus-Mercury is within 50:51 of a perfect fifth, 2:3. Only the diverging proportion of Earth and Venus, 719:1018, is far from any harmony.[19]

One must concede that these results seem worth reporting. Only about 36 percent of the proportions between pairs of randomly distributed distances should be within 50:51 of one of the seven harmonies, neglecting octaves,[20] but nine of the ten converging and diverging proportions among the distances in table 9.1 are this close to a harmony.

---

[19] The question arises why Kepler uses nonharmonic proportions such as 20:27 and 160:243—both of which appear twice in the table—to express the ratios among the distances in the table. These two proportions are among the "adulterated" or imperfect consonances given in book 3, chap. 12. They are, respectively, a fourth plus a comma and a fifth plus a comma.

[20] This estimate is based on a simulation using a pseudo–random number generator, with a uniform distribution for both numerator and denominator.

Encouraged but not satisfied—for the diverging distances of Earth and Venus are decidedly not harmonic—Kepler calculates the length of the diurnal arcs for each planet at aphelion and perihelion, multiplying the angular measure of the diurnal arcs by the mean radius of the orbit. The results, shown in table 9.2, are in effect the extreme linear velocities of the planets.

The "diurnal motions" in column 4 are obtained by converting the angular motions in column 2 to seconds and multiplying by the mean distances in column 3, where Earth's distance is taken as 1. Thus Saturn at aphelion moves 1'53", or 113", per day. Multiplying 113 by 9.510 yields 1,075, Saturn's diurnal motion at aphelion.

Kepler points out that the "proper" proportions that can be calculated from the diurnal motions in column 4 of table 9.2 are simply the inverses of the proper proportions in table 9.1, because his distance law guarantees that for any one planet, the diurnal motion is inversely proportional to distance from the Sun. The converging and diverging proportions of neighboring planets are different in table 9.2. Kepler does not present them in the text, asserting only that they are much less harmonic than before.[21]

Actually, he conjectures, it is not plausible that the Creator would have constructed the harmonies from actual paths traveled by planets in some given time. "For what purpose would there be harmonies among the paths; indeed who would perceive these harmonies?"[22] Celestial harmonies could not be audible, since there are no sounds in the heavens, so Kepler assumes that the harmonies are to be perceived visually. The calculations required to determine the planets' daily path lengths—that is, their velocities—are, he thinks, too lengthy to be carried out by any kind of natural instinct such as might be thought to perceive the planets' motions and enjoy them.

Considering everything, Kepler concluded—and rightly so, he advises the reader—that heavenly harmonies are to be sought in the apparent arcs of the planetary motions, as they would be seen from the Sun, which is the one distinguished place in the solar system, and the source of the planets' motion. He constructs a table (table 9.3) of

---

[21] *G.W.* 6.311.20–22. Recalculating, we find that no single one of the ten extreme proportions of velocities is as far from a harmony as the diverging proportion of distances was for Earth and Venus. Four of the ten new proportions, however, are close to no consonant interval at all, and the errors between the proportions and the nearest *concinna* are on the whole larger than the errors in table 9.1.

[22] *G.W.* 6.311.27–28.

TABLE 9.2.
Extreme Linear Velocities

1	2	3	4
Motion	Diurnal Minutes Seconds	Mean Distance	Diurnal motion
Saturn A	1'53"		1,075
		9,510	
Saturn P	2' 7"		1,208
Jupiter A	4'44"		1,477
		5,200	
Jupiter P	5'15"		1,638
Mars A	28'44"		2,627
		1,524	
Mars P	34'34"		3,161
Earth A	58' 6"		3,486
		1,000	
Earth P	60'13"		3,613
Venus A	95'29"		4,148
		724	
Venus P	96'50"		4,207
Mercury A	201' 0"		4,680
		388	
Mercury P	307' 3"		7,148

apparent motions that will serve as the basis for the remainder of the *Harmonice mundi.*

In table 9.3 the apparent diurnal motions in column 4 are "observed"—meaning, as often in Kepler's speculative writings, that they are calculated from astronomical theory. Those in column 5 are harmonic, agreeing exactly with the proper proportions at the right. To obtain the harmonic motions, Kepler has adjusted one of the extreme motions for each planet—the perihelial motion for the inferior planets, the aphelial motion for Earth and the superior planets—so that the proper proportion for the planet exactly equals the given

TABLE 9.3.
Apparent Extreme Planetary Motions

1	2	3	4	5	6	7
Harmonies of paired planets		Apparent diurnal motions		Closest proper harmonies of single planets		
Diverging	Converging	Planet, apse	Minutes, seconds	Minutes, seconds		
		Saturn A	1'46"	1'48"	$\frac{4}{5}$	Major third
		Saturn P	2'15"	2'15"		
$\frac{1}{3}$	$\frac{1}{2}$					
		Jupiter A	4'30"	4'35"	$\frac{5}{6}$	Minor third
		Jupiter P	5'30"	5'30"		
$\frac{1}{8}$	$\frac{5}{24}$					
		Mars A	26'14"	25'21"	$\frac{2}{3}$	Fifth
		Mars P	38' 1"	38' 1"		
$\frac{5}{12}$	$\frac{2}{3}$					
		Earth A	57' 3"	57'28"	$\frac{15}{16}$	Semitone
		Earth P	61'18"	61'18"		
$\frac{3}{5}$	$\frac{5}{8}$					
		Venus A	94'50"	94'50"	$\frac{24}{25}$	Diesis
		Venus P	97'37"	98'47"		
$\frac{1}{4}$	$\frac{3}{5}$					
		Mercury A	164' 0"	164' 0"	$\frac{5}{12}$	Octave + minor third
		Mercury P	384' 0"	394' 0"		

interval. He displays these adjusted motions in column 5 to show how nearly harmonic the actual proper proportions are.

Indeed they are. Venus is still dissonant, for its observed proper proportion at 34:35 is smaller than even the smallest *concinna*, the diesis. The error for the perfect fifth of Mars is 29:30, but all the other approximations are quite close.[23] And the proper proportions are not only melodic intervals, *concinna*, but actual consonances, excepting only those for Venus and the Earth. Even the Moon chimes in to the chorus of proper harmonies. By allowing for the lunar variation[24] Kepler shows that its actual greatest and least apparent velocities are in the proportion 3:4, a perfect fourth—which is nowhere else to be found among the apparent motions.

The "Observed" Harmonies between Adjacent Planets

Kepler turns to the extreme harmonies observed among pairs of adjacent planets, those given by the converging and diverging proportions of their apparent motions. The rough results appear in the leftmost columns of table 9.3. Kepler works through the calculation of proportions in the text, emphasizing, curiously enough, the imperfection with which these harmonies are observed. The converging proportions of the superior planets are the most pleasing: Saturn and Jupiter make exactly an octave, whereas Jupiter and Mars form a quite credible two-octaves-plus-minor-third, and Mars and the Earth a near-perfect fifth. The remaining extreme proportions do not agree very exactly with their nearest harmonies. Worst of all is the proportion of the diverging motions of Jupiter and Mars, which exceeds three octaves by something between a diesis and a semitone, an excruciating interval that must surely distress any sensitive and musically inclined soul in the Sun.

Kepler has already remarked that the errors for the proper proportions of Mars and Mercury are "about 34:35 or 35:36"—about the same size, in fact, as the proper proportion of Venus. Here he points out that

[23] The worst of the others is Mercury, for which Kepler admits an error of 38:39, although the error computed from table 9.3 is only about 40:41.

[24] This is surely a legitimate adjustment, for the variation describes a real change in the Moon's velocity around the Earth. The other lunar inequalities are irrelevant: the evection does not affect lunar velocity at the apsides, and Kepler believes that the annual equation pertains to the equation of time rather than the lunar motion (Stephenson, *Kepler's Physical Astronomy* [New York, 1987], pp. 197–201).

three of the extreme proportions (the diverging proportions of Mars-Earth and Earth-Venus, and the converging proportion of Earth-Venus) differ from the nearest harmony by very nearly this same proper proportion of Venus. Moreover, if the aphelial motion of Saturn is increased by a mere one second of apparent arc, the error in the diverging motions of Saturn and Jupiter would be just about this same amount.

Kepler is not rounding these proportions creatively but rather directing his attentive reader to the way that this particular very small interval seems to be implicated in the problematic harmonies of planetary motion. He does not think that the errors prevent the overall effect produced by the extreme motions of adjacent planets from being musically pleasing. If pairs of strings are tuned to the converging and diverging motions, the imperfection in the harmonies would not easily be heard, he says, except in the single case of the diverging motions of Jupiter and Mars. In fact, he intends to show that there are reasons for all the imperfections observed in the harmony of the planetary motions—reasons arising from internal constraints within the overall design.

Before leading his reader in the pursuit of these constraints, Kepler looks at another type of harmony among motions of adjacent planets, comparing what he calls motions *eiusdem plagae*. This is his term for pairs of either aphelial or perihelial motions of different planets. The phrase means something like "in the same region," describing motions that are both either outermost or innermost. Rather than calculate them from the data in table 9.3, he derives them from the previously obtained extreme and proper proportions, using the identities

$$A_s{:}A_i = A_s{:}P_s + P_s{:}A_i$$
$$P_s{:}P_i = P_s{:}A_i + A_i{:}P_i$$

where A_s, A_i are the apparent aphelial motions of the outer and inner planets, and P_s, P_i the apparent perihelial motions. The terms on the right are known, being the proper proportions and the converging proportions of the two planets. (Kepler chooses to work from the converging proportions rather than make analogous calculations from the less harmonious diverging proportions.)

Not surprisingly, most of the proportions *eiusdem plagae*, thus calculated, are as close to being harmonious as the proper and extreme proportions used to calculate them. For a few instances (the perihelial proportion of Jupiter and Mars, the aphelial proportion of Mars and Earth, and both proportions of Venus with Mercury) the errors accumulate, so that the proportions *eiusdem plagae* are in those instances less harmonious than the converging proportions are.

Between the Earth and Venus, however, Kepler greatly reduces the harmonic errors—errors that, as he has remarked, roughly equal the proper proportion of Venus—by using the above identities to switch from extreme proportions to proportions *eiusdem plagae*.[25] One can easily see how this works:

$$A_e{:}A_v = A_e{:}P_v - A_v{:}P_v$$
$$= (3{:}5 + 34{:}35) - 34{:}35$$
$$= 3{:}5$$
$$P_e{:}P_v = P_e{:}A_v + A_v{:}P_v$$
$$= (5{:}8 - 29{:}30) + 34{:}35$$
$$\approx 5{:}8$$

To summarize all this, Kepler has found, by examining the "observed" harmonies among the motions of adjacent planets, reasonably precise harmonies between the converging motions of Saturn and Jupiter, Jupiter and Mars, Mars and Earth, and Venus and Mercury; between the diverging motions of Venus and Mercury; between the aphelial motions of Earth and Venus; and between the perihelial motions of Mars and Earth, Earth and Venus, and Venus and Mercury. The small discrepancies in other pairs of motions, he argues, can easily be "swallowed" without impugning the astronomy he has constructed from Tycho's observations—especially because the largest errors occur in the proportions involving Venus and Mercury, whose motions are not very well established.

Before turning to the musical interpretation of these harmonic proportions, Kepler points out four things he thinks his reader should remember:

1. The least accurate of the extreme harmonies occur where the polyhedral hypothesis is most accurate, in the diverging interval between Jupiter and Mars.

2. The small proper proportion of Venus turns up several places, for some reason, as an error in the observed extreme proportions.

3. The best harmonies among the superior planets are between converging motions, while the best harmonies among inferior planets seem to be between motions *eiusdem plagae*.

---

[25] Kepler actually works out the first of the adjustments shown above using the converging proportion plus the proper proportion of the Earth, which yields 21,235:35,328, extremely close to a perfect fifth.

4. The aphelial motions of Saturn and Earth make almost exactly five octaves.

All four of these points figure prominently in the detailed discussions to come.

CHAPTER 5: THE MUSIC OF THE PLANETARY MOTIONS

Harmony in the motion of one planet—the harmonic proportion between its aphelial and its perihelial motions—cannot be perceived in any single moment, simply because the planet cannot at the same time be both at aphelion and at perihelion. Harmonies between different planets, on the other hand, can be perceived at those (rare) moments when one planet is at aphelion while its outer neighbor is at perihelion. Harmony in the motion of a single planet, then, is like a song for one voice, whereas harmony among different planets is polyphonic.

Kepler turns first to sequential, or melodic, harmonies. He knows that the notes played by extreme motions of individual planets sound pleasingly together in pairs. He has never, he says—exaggerating a bit here—failed to find harmonic proportions in all the comparisons he has made. "Unless all of them are fit into a single scale," he continues, "it could have easily happened (as here and there it did happen, by force of necessity) that several dissonances would occur." By this he means that unfortunate combinations of consonant intervals produce dissonances, if all the notes are not tuned into a single scale. There is no room in a scale, for example, for a major third on top of a major sixth; the combination of these consonant intervals produces the painful dissonance 12:25.[26] Certainly Kepler has pointed out no such dissonant intervals among the extreme motions in chapter 4.

At any rate, he turns in chapter 5 to the agreement between the apparent planetary motions and the notes of a scale. For his initial attempts to fit the planetary motions into a scale, he warns, it is necessary to disregard errors less than a semitone; the causes of those will be

---

[26] *G.W.* 6.317.23–25. One cannot help noticing that the parenthetical concession "as here and there it did happen" invalidates this argument. Chap. 5 was evidently drafted under the belief that there were no dissonances, and revised after the grueling work of the *posteriores rationes* of chap. 9 had convinced Kepler that some dissonances were logically necessary.

TABLE 9.4.
Reduced Extreme Motions

1 Motion	2 Octaves reduced	3 Divisor	4 Reduced motion
Mercury P	7	128	3'0"
Mercury A	6	64	2'34" −
Venus P	5	32	3'3" +
Venus A	5	32	2'58" −
Earth P	5	32	1'55" −
Earth A	5	32	1'47" −
Mars P	4	16	2'23" −
Mars A	3	8	3'17" −
Jupiter P	1	2	2'45"
Jupiter A	1	2	2'15"
Saturn P	0	1	2'15"
Saturn A	0	1	1'46"

explained later.[27] In order to compare the "notes" of the different planets, some of which moved much faster than others, he divides their extreme motions by two as many times as is necessary to make them comparable. Musically, he is shifting the high notes down by octaves until they are within an octave of the lowest note, which is Saturn's aphelial motion. Table 9.4 shows the extreme motions reduced into that octave. (The + and − signs after some of the motions in column 4 merely indicate which quotients in that column have been rounded.)

The next step is to express the reduced motions in column 4 as musical notes. He does this at two different tuning levels. First, he places Saturn's aphelial 1'46" on the fundamental G. The Earth's aph-

[27] Ibid., 6.317.32–34.

elial motion, five octaves higher, is at 1'47", and "who has dared to argue about one second in Saturn's aphelial motion?" The other motions are assigned to notes according to their proportion to one another, using the Earth's 1'47" as the standard. Saturn's motion at perihelion and Jupiter's at aphelion are both about a major third above G, since ⅘ times 1'47" is 2'13¾", so Kepler assigns both those motions to B. Jupiter's B is two octaves higher, but that does not affect the harmonies. Placing all the extreme motions as well as he can, he displays the notes as in figure 9.4.

The two motions at G enable Kepler to place one of them at the bottom of the octave and one at the top. The aphelial motion of Mars falls a bit closer to F♯ than to F; Kepler observes that in music F♯ is often used in place of F. All the notes of the scale are present except A; and all the extreme motions fit into the scale except three. Of those three motions, Mercury's aphelial at least corresponds to a note; that note, however, is C♯, which is not a part of the scale from G.

Kepler also defines a second tuning, wherein G corresponds to the 2'15" of Saturn's *perihelial* motion. The scale in figure 9.5 is mollis. A single note is again absent, here F; and all the extreme motions are heard except three.

Let us collate these notes into a single table, arranging them in order of pitch. In table 9.5 the extreme motions appear in column 1 for reference. (The motions for the highest three notes have been raised by an octave, to account for their appearance at the top of the scale on Kepler's staves.) Columns 3 and 4 give the notes in the tuning from figure 9.4, along with the motions that would correspond exactly to

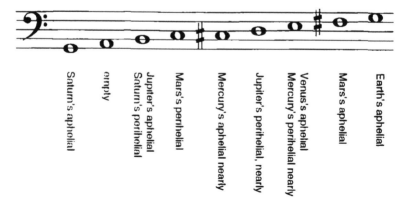

FIGURE 9.4. Extreme motions as notes

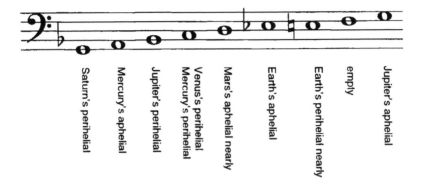

FIGURE 9.5. Extreme motions as notes, second tuning

those notes. Columns 5 and 6 do the same for the tuning in figure 9.5. (Note, as a detail, that the word *nearly, fere,* modifies the actual motions, not the pitches, since that is how Kepler uses the word in figures 9.4 and 9.5.)

The motions to which the scales are tuned are in boldface. Two such motions appear for the first tuning; Kepler describes it as tuned to Saturn's aphelial motion of 1'46" but calculates the pitches in column 4 from Earth's aphelial motion of 1'47", or 3'34" in the higher octave.

By moving the fundamental G from Saturn's aphelial motion to its perihelial motion, Kepler has shifted the tuning of all the (approximate) notes played by the planetary motions, by about a major third. Since the notes of his scale are not equally spaced, some of the motions are closer to notes in the scale tuned to Saturn's aphelial motion, and some to notes in the scale tuned to its perihelial motion. The ambiguity of note assignments in this unequally tempered scale is most visible for the perihelial motion of Venus, which Kepler has reduced by five octaves (a factor of 32) to 3'3". This makes a proportion of 3:4 + 60:61 with the perihelial motion of Saturn, quite close to a perfect fourth, and so this motion appears in figure 9.5 at C, a fourth above G. It makes a proportion of 3:5 + 28:29 with the aphelial motion of Saturn, nearly a diesis away from the major sixth, in figure 9.4. Hence Kepler does not assign it to E, the note that is a major sixth above G.

TABLE 9.5.
Planetary Motions in Two Tunings

1	2	3	4	5	6
		Saturn's aphelial motion = G		Saturn's perihelial motion = G	
	Extreme motion				
4'30"	Jupiter's aphelial	(B in lower octave)		G	4'30"
3'49"−	Earth's perihelial	—		nearly E	3'45"
3'34"−	Earth's aphelial	**G**	**3'34"**	D♯	3'36"
3'17"−	Mars's aphelial	F♯	3'21"	nearly D	3'23"
3' 3"+	Venus's perihelial	—		C	3' 0"
3' 0"	Mercury's perihelial	nearly E	2'58"	C	3' 0"
2'58"	Venus's aphelial	E	2'58"	—	
2'45"	Jupiter's perihelial	nearly D	2'41"	B♭	2'42"
2'34"−	Mercury's aphelial	nearly C♯	2'30"	A	2'32"
2'23"−	Mars's perihelial	C	2'23"	—	
2'15"	Jupiter's aphelial	B	2'14"	(G in higher octave)	
2'15"	Saturn's perihelial	B	2'14"	**G**	**2'15"**
1'46"	Saturn's aphelial	**G**	**1'46"**	—	

The Genera of the Scales

Kepler does not display notes indiscriminately on his staves. He is no longer arguing merely that harmonic proportions exist among the planetary motions; he is exhibiting scales—scales, moreover, from each genus of harmony. The scale based on the aphelial motion of Saturn includes B and E, the durus third and sixth, and belongs consequently to cantus durus. The scale based on the perihelial motion of Saturn includes B♭ and E♭ (written Dρ, i.e., D♯), and so belongs to cantus mollis.[28] The existence of scales of both genera is of great

---

[28] The Earth's perihelial motion 3'49" intrudes at E into the mollis scale. Kepler

significance in Kepler's understanding of the heavens and is therefore essential to an understanding of the *Harmonice mundi*, particularly the labyrinths of chapter 9.

It is at first surprising to notice that some of the notes omitted from table 9.5 would fit rather well into these scales. Evidently Kepler chooses to leave them out. The aphelial motion of Venus rightfully belongs at C in the second scale, for this motion is actually closer to a perfect fourth from G than is the *perihelial* motion of Venus, which Kepler does place at C. He has his own reasons, discussed below in some detail, for preferring the perihelial motion of Venus to its aphelial motion in the mollis scale.

Of the other four extreme motions missing from one scale or the other, three actually do fit some note moderately well. In the first scale, the perihelial motion of Earth is five octaves and a limma, plus 35:36, above the aphelial motion of Saturn, thus a not very well tuned G♯. (The first interval above the fundamental—the interval from G to G♯—is a limma in Kepler's system of the octave.[29]) In the second scale, the perihelial motion of Mars is almost exactly four octaves and a limma above the perihelial motion of Saturn, quite a well tuned G♯. The note G♯, however, is dissonant in both genera of scales, so it is not surprising that Kepler omits it. The other extreme motion that Kepler omits is the aphelial motion of Saturn, only 53:54 less than being a minor sixth above its perihelial motion and hence sufficiently close to the note D♯. Kepler already has an D♯ in the mollis scale and apparently feels no need to add another by mentioning Saturn's note five octaves lower.

Kepler summarizes his efforts with a picture (figure 9.6) and a few words.

Even when one knows that Kepler is trying to show how well the notes in figures 9.4 and 9.5 fit into scales, this figure is not entirely clear. The bottom half depicts the durus and mollis scales, and the top half, those scales as expressed in the heavens. The numbers below each note show the relative length of string that would sound that note. These numbers, which first appeared in chapter 8 of book 3,[30] are the smallest integers that exactly express the proportions among

---

erroneously computed 3′50″ as the motion corresponding to E in that tuning (*G.W.* 6.319.20) and perhaps thought that the Earth's perihelial motion was too close to ignore. In fact a major sixth above 2′15″ is exactly 3′45″.

[29] Ibid., 6.148.12.

[30] Ibid., 6.142.

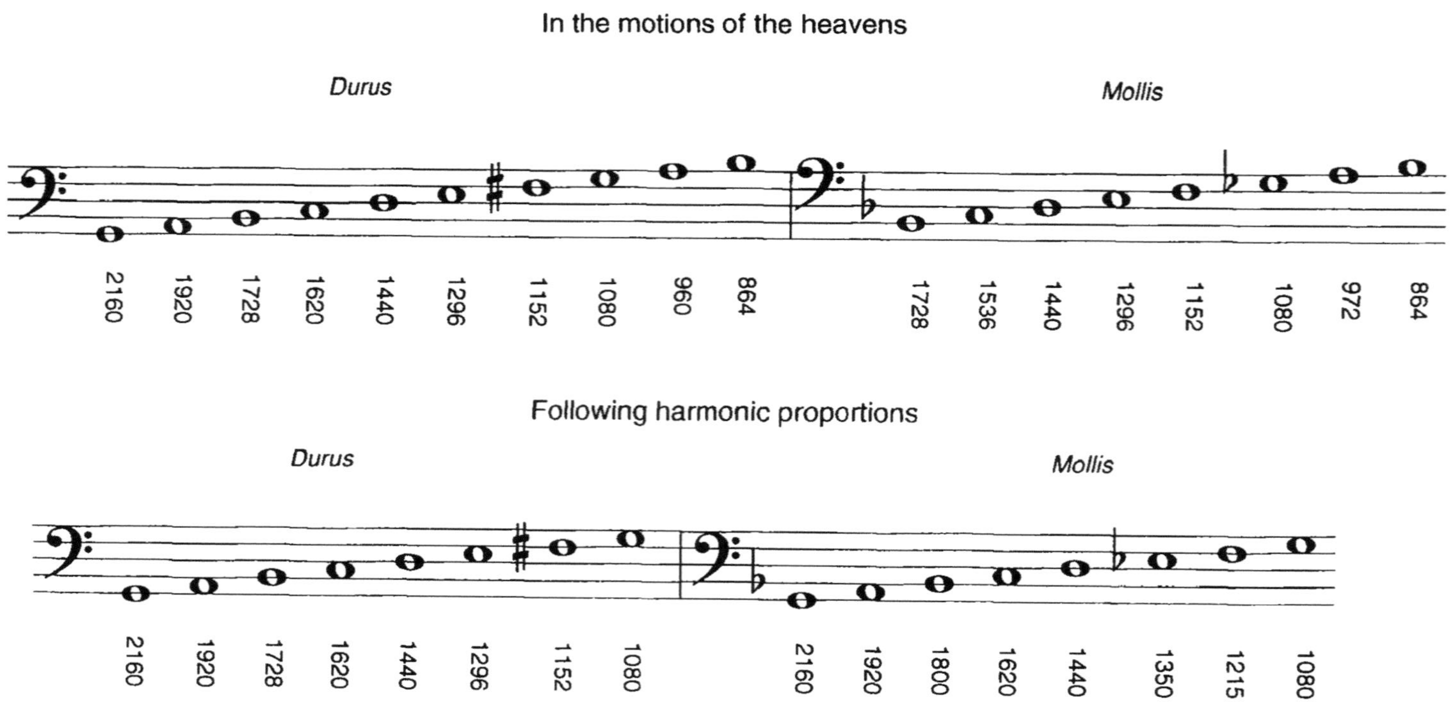

In the motions of the heavens

Mollis

864
972
1080
1152
1296
1440
1536
1728

Durus

864
960
1080
1152
1296
1440
1620
1728
1920
2160

Following harmonic proportions

Mollis

1080
1215
1350
1440
1620
1800
1920
2160

Durus

1080
1152
1296
1440
1620
1728
1920
2160

FIGURE 9.6. Genera of music in the heavens

the twelve notes in Kepler's system of the octave. Since they represent string lengths, larger numbers correspond to lower notes.

The upper half of the figure represents the durus and mollis scales, as those scales are expressed in the heavenly motions. It contains most of the notes from figures 9.4 and 9.5, omitting those that do not fit well into the scale. To the left are the notes of the durus scale of figure 9.4 (omitting the C♯ that is not a legitimate part of the scale), extended at the top by two extra notes and labeled with numbers representing the appropriate string lengths. To the right are the notes of the mollis scale of figure 9.5 (omitting the E♮ not a part of the scale), labeled with string lengths. This mollis scale is displayed on a shifted clef: the clef signature places F on the fifth (top) line, so that this is indeed a mollis scale from G like that in figure 9.5 and like the one immediately below it in the lower part of figure 9.6.

A more detailed examination of the figure reveals that it is almost annoyingly complex. Readers whose interest in these scales has been sated may wish to skip ahead to the section on alternate tunings.

The string lengths for the mollis scale of heavenly motions (at the top right) obviously differ from those for the mollis scale of harmonic divisions (at the bottom right) but agree with the string lengths for the same planetary motions in the durus scale to their left. The fundamental G in the mollis scale (top right) has a length of 1,728 because in the heavens it is expressed by the perihelial motion of Saturn, a motion that in the durus scale (top left) plays B and is assigned a length of 1,728. The B♭ in the mollis scale similarly has a length of 1,440 because it is expressed by the perihelial motion of Jupiter, which plays D in the durus scale and has been assigned a length of 1,440 on the durus side of the figure. To abstract a rule from this, the string lengths for the mollis scale of heavenly motions are determined by the motions and not by the notes. This will, I hope, become clearer in the table that follows.

Kepler has attempted to clarify the relationships by shifting the clef signature at the top right. Saturn's perihelial motion, with its length of 1,728, appears on the second line (from the bottom) of the staff to the left and remains on the second line of the staff to the right. Jupiter's perihelial motion likewise remains on the third line of the staff. Mercury's perihelial motion remains on the third space from the bottom; Mars's aphelial motion on the fourth line; and Earth's aphelial motion on the fourth space. Table 9.6 gathers together the intricate relationships underlying the staves Kepler printed in chapter 5 of book 5.

TABLE 9.6.
Scales in the Motions of the Heavens

1	2		3 Length of string	4 Durus note	5 Mollis note
Position on staff	Motion				
Space above staff	same as second line		864	B	G
Top line	None		972	not in scale	F
	None		960	A	not in scale
Top space	Earth A	3'34"	1,080	G	E♭
Fourth line from bottom	Mars A	3'17"	1,152	F♯	D
	Venus A	2'58"	1,296	E	not used
Third space	Mercury P	3'00"		E	C
	Venus P	3'03"		not used	C
Middle line	Jupiter P	2'45"	1,440	D	B♭
Second space	Mars P	2'23"	1,620	C	not in scale
	Mercury A	2'34"	1,536	not in scale	A
Second line from bottom	Saturn P	2'15"	1,728	B	G
	Jupiter A	2'15"		B	G
Bottom space	None		1,920	A	
Bottom line	Saturn A	1'46"	2,160	G	

Each line and space on the upper staff of figure 9.6 corresponds to one amount of apparent motion; to a string of a definite length, perhaps differing a bit between the durus and mollis sides; and—because of the shifted clef signature—to two different notes, one for each genus. Two positions on the staff are assigned different string lengths on the durus and mollis sides of the figure. (These slightly different tunings are due to the shifted clef signature.) The first is the fifth or top line. At the left, A on the top line is a major tone above G in the fourth space, whereas at the right F on the fifth line is only a minor tone above the E♭ in the fourth space.[31] Hence the top line corresponds to a string length of 960 at the left but of 972 at the right, making a proportion of a comma, 80:81—which is the difference between the major and minor tones. The second exception is the second space from the bottom of the staff. To the left, C in the second space is a semitone above B in the second line, and a major tone below D in the middle line. To the right, A in the second space is a major tone above G in the second line and a semitone below B♭ in the third line. The corresponding string lengths are 1,620 for C at the left and 1,536 for A at the right, making the proportion of a limma, 128:135. This, of course, is the difference between the major tone and the semitone.[32]

The staves labeled "in the motions of the heavens" thus summarize the harmonic proportions Kepler has found in the extreme motions of the planets. It is not, unfortunately, a summary that becomes clearer the more carefully one looks at it, until one has looked at it very carefully indeed. Placement of clef signatures on the staff was more flexible in Kepler's time, so the shifting of the signature to align the durus and mollis motions no doubt confused his contemporaries less than us. His main point, at any rate, is simple: the extreme planetary motions yield most of the proportions among the notes of the durus scale and also most of the proportions among the notes of the mollis scale. Kepler makes a strong claim in chapter 5 that the apparent planetary motions embody both types of musical scale.

Alternate Tunings

Before moving on, Kepler briefly indicates a way in which scales (of both genera) can be based on the motions of Venus as anchors, rather than those of Saturn. If the aphelial motion of Venus is assigned *exactly* to E, the note to which it closely approaches in the tuning based on Saturn's aphelial motion (figure 9.4), then the other motions given in column 3 of table 9.5 can still be identified with the same notes in the

[31] See tables 8.1 and 8.3 for definitions of the major and minor tones.

[32] Within the portion of the staff used in both scales, the other intervals between lines and spaces agree. From line 3 to space 3 is a minor tone, from space 3 to line 4 a major tone, and from line 4 to space 4 a semitone, on both clefs. The intervals between the first line, the first space, and the second line would differ also, but those intervals are below the mollis scale in Kepler's figure.

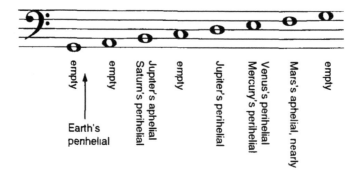

FIGURE 9.7. Extreme motions as notes, third tuning

durus scale. Tuning the scale to Venus does not change any of the notes in the durus scale enough to matter.

If, however, the perihelial motion of Venus—instead of the aphelial—is shifted to E, one obtains a variant form of the durus scale, which Kepler describes verbally.[33] This scale is shown in figure 9.7.

Kepler points out that the perihelial motion of Earth, a minor sixth below Venus's perihelial E, does not fit (*non quadrat*) here. This is because a minor sixth below E is exactly a diesis above G and consequently lies between G and G♯ (as one no doubt recalls from table 8.2, G♯ is a limma above G). Kepler needs to include these two perihelial motions in the same scale, for reasons that will become apparent when we look at chapter 9, and so he rejects this alternate durus scale.

The second durus scale can be converted into a mollis scale by shifting the perihelial motion of Venus down by precisely the diesis that separated Earth's perihelial motion from G. Kepler mentions this second mollis scale only in passing in chapter 5, not so much describing as implying it. It is anchored by fixing the perihelial motion of Venus, 3′3″, at D♯ (a diesis below E). This note at D♯ converts the scale to the mollis type. Mercury's perihelial[34] motion, two octaves higher than that of Venus, is also on D♯. The Earth's perihelial motion is a minor sixth lower than D♯, coming to rest exactly on G. The aphelial motion of Mercury is a major sixth higher than Venus's D♯, hence on C.

[33] *G.W.* 6.320–321.12. If Kepler had discussed this tuning further, he might have included the note G, since a major sixth below 3′3″ is 1′50″, close to the Earth's aphelial motion at 1′47″; and the note C, since a major third below 3′3″ is 2′26″, close to Mars's perihelial motion at 2′23″.

[34] Read *perihelius Mercurii* instead of *aphelius Mercurii* on p. 206, line 14, of the 1619 edition, or in *G.W.* 6.321.12–13.

TABLE 9.7.
Alternate Mollis Scale

1	2	3	
		Venus's perihelial	
	Extreme motion	*motion = D♯*	
3′ 3″+	**Venus perihelial**	D♯	3′ 3″
3′ 0″	Mercury perihelial	D♯	3′ 3″
2′34″−	Mercury aphelial	C	2′32″
1′55″	Earth perihelial	G	1′54″

In table 9.7 we display these four motions, in column 1, and the notes to which they correspond, in column 3. Only the notes cited by Kepler are shown, although some of the other extreme motions do correspond to notes, not all of which are a part of the mollis scale.[35]

The alternate mollis scale to which Kepler alluded in ending chapter 5 would thus have resembled figure 9.8. The figure includes two motions not mentioned in Kepler's text, because they both fit into the scale on the note B♭.

Closing chapter 5, Kepler advises his reader not to make too much of these results, which are after all only approximate. It is enough that the presence of major and minor scales in the heavens has been noted. The causes not only of the harmonies but also of the remaining errors will be made clear, he promises, in the splendid demonstrations of chapter 9.

CHAPTER 6: HEAVENLY MODES

Perhaps it is the variety of the scales to which Kepler attached the planetary motions in chapter 5 that has made the otherwise unremarkable chapter 6 confusing for historians. This short chapter is in fact largely irrelevant to the main argument of the book, but it contains

---

[35] *G.W.* 6.321.12–17. A fourth down from 3′3″ is 2′17″, comfortably close to Jupiter's aphelial motion and Saturn's perihelial, so those two motions are in the scale at B♭. Other motions could be placed at C♯ (perihelial of Jupiter), at E (aphelial of Mars), and at F♯ (aphelial of Saturn and Earth); but those notes are not in the mollis scale.

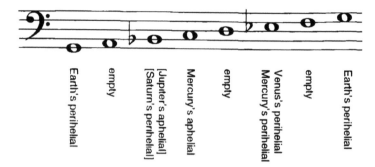

FIGURE 9.8. Extreme motions as notes, fourth tuning

some musical staves (figure 9.9) that seem to summarize the celestial music of the *Harmonice mundi*. Kepler by no means intends for them to do so. For him the point of chapter 6 is simply to allocate each of the planets to a mode (modus or tonus) among those used in ecclesiastical music. As a basis for this discussion he displays the proper intervals of all the planets on musical staves, choosing freely from the various tunings he used in the previous chapter. The intermediate notes, he explains, are not articulated separately but are passed through as the planet moves from its lowest note at aphelion to its highest at perihelion.

The first problems here are the multiplicity of clef signatures and Kepler's free placement of them. Saturn is on the familiar bass or F clef, climbing from G up to B and back. Jupiter is on an F clef with F moved to the top line of the staff, sometimes known as a sub–bass clef; hence Jupiter's chant lies between G and B♭. The staff for Mars uses a tenor clef, placing C at the fourth line from the bottom, so its note varies from F to C and back. Earth's staff uses a G clef with G on the middle line; Earth thus moves between G and A♭. Venus is on a G clef in its usual position (the "treble clef"), and stays always at E.[36] Mercury is on a C clef with C at the second line from the bottom, and thus its long

[36] Dickreiter, *Der Musiktheoretiker Johannes Kepler. Neue Heidelberger Studien zur Musikwissenschaft, Band 5* (Bern and Munich, 1973), p. 216, n. 20, says that Venus should move between E and E♯, and that Kepler ignores this to suppress a dissonance between Venus and Mercury. In fact Kepler never uses the note E♯. The interval between E and F is a semitone in Kepler's system, more than twice the proper interval of the motions of Venus. A more general problem with Dickreiter's assertion is that intervals between different planets are not relevant in chap. 6, which merely assigns modes to the individual planets.

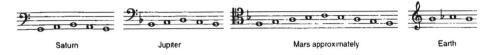

Saturn   Jupiter   Mars approximately   Earth

Venus   Mercury   The moon also has a place

FIGURE 9.9. Proper intervals of the planets

glissando slides between A and C a minor tenth higher. And the Moon, on a shifted G clef, wanders between G and C.

Because these figures *seem* to summarize a great part of Kepler's theory—although they do not—they have assumed an importance all out of proportion to Kepler's use of them in assigning planets to modes. Along the way not a few historians have felt the need to correct them, with Mercury suffering the worst treatment. Frisch, in his edition of the *Harmonice mundi*, silently shifted the clef sign for Mercury's staff to make that planet move from C to E a major tenth higher—which is somewhat surprising, since Kepler clearly says that this interval is a minor tenth.[37] In fact, no emendations are needed to these staves or to the positions of the clef signs. Fortunately, Caspar, when he edited the *Harmonice mundi* for the *Gesammelte Werke*, left the figure as Kepler published it.

The different staves of figure 9.9 correspond to the two principal tunings from chapter 5. The aphelial motions of Saturn, Venus, and Earth are tuned to G, E, and G, as in Kepler's primary durus tuning, figure 9.4. The motions of Mars are placed "approximately" from F to C, because its proper interval is about a fourth, and in the discussion

---

[37] Frisch, *O.O.*, 5.294. Apelt, *Johann Keppler's Astronomische Weltansicht* (Leipzig, 1849), p. 93, perhaps originated this correction. Dreyer, *A History of Astronomy* (New York, 1953), p. 408, and Eliot Carter, on p. 1039 of Kepler's *Harmonies of the World,* accepted it, perhaps on the authority of Frisch's edition. Koyré, *The Astronomical Revolution* (Ithaca, 1973), p. 337, and Walker, "Kepler's Celestial Music," in *Studies in Musical Science in the Late Renaissance* (London, 1978), p. 59 and n. 75, followed along with less justification. Dickreiter, *Der Musiktheoretiker Johannes Kepler. Neue Heidelberger Studien zur Musikwissenschaft, Band 5* (Bern and Munich, 1973), p. 216, n. 21, narrated this much but then concluded incorrectly that Mercury's interval really should be from C♯ to E, the notes near which its motions were placed in fig. 9.4 above.

leading to figure 9.4 its aphelial motion is somewhere between F and F♯. Those of Jupiter and Mercury are tuned to G and A as in the primary mollis tuning in figure 9.5. This use of different tunings for the different planets should not be confusing, for in this chapter Kepler listens to the planets singly. The proper planetary intervals are all that matters for the purposes of chapter 6; Kepler has simply chosen a note that expresses the aphelial motion of each planet in one of the primary tunings and based its up-and-down melody on that note. The melodies of the different planets are not intended to be comparable in pitch.

From the scale fragments in figure 9.9 Kepler assigns each planet to one of the musical modes,[38] on the basis of a general theory he has developed in book 3 of the *Harmonice mundi.* That theory distinguishes the possible ways in which the twelve intervals in the system of the octave (table 8.2) can be combined to form a scale of eight notes separated by major or minor tones or semitones. The durus and mollis scales shown in table 8.3 are the most common ways of doing this but by no means the only ways. Kepler methodically enumerated all possible scales in chapter 14 of book 3, by starting from each of the twelve notes in the system of the octave and combining semitones with adjacent limmas or dieses to form major or minor tones.

Most of the "scales" obtained by such combinations are disagreeable (*inconcinna*) according to one or more of ten laws propounded in book 3, chapter 13. The eighth law, for example, states that the first three intervals at the bottom of the octave cannot all be tones—in other words, that at least one of them must be a semitone. By the third law, two semitones are not allowed in the lowest fifth of the octave. Hence exactly one of the first three intervals must be a semitone, whereas the other two must be major or minor tones.[39] From these laws Kepler determined that precisely fourteen acceptable types of scale (twenty-four, if he counts variants in the top three notes) can be carved out of the twelve notes in his system of the octave.[40]

He continued in book 3 by reconciling this general theory of scales with the eight common modes (*Tonos octo vulgares*) known as the

---

[38] The term *mode* carried several meanings in ancient music theory, some of which remain unclear; see Palisca, *Humanism in Italian Renaissance Musical Thought,* (New Haven, 1985), pp. 39–47, and Randel, ed., *New Harvard Dictionary of Music* (Cambridge, Mass., 1986) s.v. "Greece."

[39] *G.W.* 6.160.33–161.34.

[40] Ibid., 6.165–70. See the discussion in Dickreiter, pp. 170–76.

ecclesiastical modes. The Dorian and Hypodorian modes both[41] correspond to one of his scales, a mollis scale based on the note G; the Phrygian and Hypophrygian modes to an unconventional scale based on E; the Lydian and Hypolydian to one based on F; and the Mixolydian and Hypomixolydian to a durus scale based on G. The other ten of his scales differ from all the traditional modes.[42]

This elaborate and highly abstract theory from book 3, which he has grudgingly reduced to account for the eight paltry ecclesiastical modes, is Kepler's standard of comparison in the sixth chapter of book 5. He applies it to very unworthy data: the extreme notes that the planets play in the "common system" of figures 9.4 and 9.5. Intermediate intervals, which are after all the essence of distinction between modes, are not even defined by the continual variation of a planet's motion between its aphelial and perihelial extremes. Even without the intermediate notes, though, Kepler contrives to assign each planet to one of the four pairs of ecclesiastical modes, as follows.

Saturn's extreme notes are G and B. These two notes alone imply a durus scale from G, hence the Mixolydian or Hypomixolydian mode, since they are the only modes associated with his durus scales. Jupiter's notes are G and B♭, evidently from a mollis scale on G. This implies for Jupiter the Dorian or Hypodorian mode. The proper interval of Mars is close to a perfect fifth, an interval that appeared in every mode; but its aphelial note in the durus scale of the common system is F (or nearly so), weakly indicating the Lydian or Hypolydian mode.

Earth takes only one step, and that a mere semitone, up its scale before coming back down to its aphelial note. Since the Phrygian and Hypophrygian are the only modes beginning with a semitone, Kepler assigns Earth to them, although its aphelial note is E in neither scale of his "common system." This Phrygian mode (which he characterized in book 3 as "plaintive, broken, and mournful") can be sung by beginning with *mi* in the *ut–re–mi–fa* syllabic notation, or solmization, of the octave, so Kepler puns in a marginal note that Earth's song is *mi–fa–mi* because of the misery and hunger, *miseria et fames*, in this world.[43]

Venus stays on a single note from aphelion to perihelion. Because

---

[41] Kepler thinks the distinction between, for example, the Dorian and Hypodorian modes to be harmonically unimportant, as such pairs of modes are based on the same scale but one using a higher or lower melodic range.

[42] *G.W.* 6.170.32–171.30.

[43] Ibid., 6.177.13–16; 6.322.27–30.

that note is E in the durus scale of the common system, Kepler assigns Venus to the same Phrygian-Hypophrygian modes as Earth. The proper interval of Mercury is greater than an octave, and neither of the notes assigned to its aphelial motion is the basis of an ecclesiastical mode. Kepler concludes therefore, agreeably to Mercury's androgynous character in astrology, that it is suited to all the modes indiscriminately.

Kepler thus is able to assign all the planets except Mercury uniquely to one of the four pairs of modes, using one or both of two lines of reasoning. First, Saturn, Jupiter, and Earth have proper intervals that indicate their modes uniquely. Second, when the planets are all placed into the "common system" of figures 9.4 and 9.5, Saturn, Jupiter, Mars, and Venus begin their motions from the characteristic lowest note of their assigned modes. Such assignments obviously are not unique; indeed Kepler himself made them differently in an earlier version of this theory.[44] In this version, the planets displayed among them all eight of the modes. This is Kepler's whole point in chapter 6 of book 5—not that the musical passages in figure 9.9 summarize the harmonies of the world but that the variety of those harmonies includes all eight of the traditional ecclesiastical modes. It is a small point but one that lends support to the analogous conclusions of neighboring chapters.

Chapter 7: Universal Harmonies

In Chapter 7 Kepler finally considers how polyphonic harmony can arise from the apparent planetary motions. Polyphonic harmony—in which multiple voices are heard at the same time—is in Kepler's opinion the crowning achievement of "modern" music. By creating polyphony musicians are expressing in analogy the innermost secrets of Creation.

Simultaneous occurrences of apparent motions in harmonic proportion can occur at any time. Whatever a planet's apparent motion, that of another planet might momentarily harmonize with it. Such harmonies are too numerous, and too fleeting, to be of any great significance. The most extraordinary moments of harmony are those involving the extreme motions: the rare events when two or more

---

[44] *O.O.* 5.385, n. 5.

planets are simultaneously in their apsides and moving in harmonic proportion to one another. Just how extraordinary these events are depends, of course, upon how narrowly one defines the moments of aphelion and perihelion. Since planetary velocities change very slowly around the apsides, Kepler can endorse events in which the planets are reasonably near their apsides rather than precisely at them. He recognizes, though, that simultaneous occurrences of even the nearly extreme motions happen at long time-intervals, particularly for the outer planets.

Harmonic proportions outside the apsides are much more numerous, hence less rare. As Saturn moves between G and B, and Jupiter between B (an octave higher) and D, harmonic intervals are possible of an octave, or of an octave plus a major or minor third, a fourth, or a fifth. The octave and the octave-plus-fifth happen only at the apsides, but the others can occur anywhere within a range of motions. An octave-plus-minor-third, for example, can occur at any point in Saturn's orbit where its note is G♯ or higher. Harmonies involving Mercury are the most numerous. Mercury's proper proportion exceeds an octave, so in passing from aphelion to perihelion—within a time interval of only forty-four days—it momentarily sings every harmonic interval at least once with each of the other planets.

Two-note harmonies of intermediate motions are thus almost a daily occurrence.[45] Three-note harmonies occur fairly often, particularly among Mercury, Earth, and Mars, all of which slide up and down their proper intervals with reasonable alacrity. Venus, whose note scarcely varies, is constrained to await passively the moments when other planets come into harmonic proportion with it; it is left out of most of their momentary chords. The attainment of a four-note harmony must therefore await the ponderous changes of Jupiter's or Saturn's motion in the bass, or else settle on the monotone of Venus. "Harmonies of four planets," Kepler writes, "begin to spread out among the centuries; those of five planets, among myriads of years."[46]

The greatest harmonies, those among all six planets, would truly be of a rarity befitting their grandeur. It might not be possible, Kepler admits, for such a "universal harmony" to happen more than once in

[45] In a letter of 1608 Kepler speculates that such fleeting moments of celestial harmony might have physical effects analogous to those of the equally fleeting moments when planets are in astrological aspect. Kepler to J. Tanckius, 12 May 1608, no. 493 in *G.W.* 16.164.426–165.436. This idea reappears in the *Epitome*, ibid., 7.512.16–21.

[46] Ibid., 6.324.5–11.

all the regular course of heavenly motions. He suspects that a past moment of universal harmony, if one could calculate it, would point out the moment of creation—the beginning of time itself. Certainly the architect of the heavens had ample reason to start the planets spinning from a universal harmony, even if he did not intend that the world should last long enough for that great harmony to recur.

Skeletons of Universal Harmony

The very possibility of universal six-note harmonies cannot be assumed; it must be demonstrated. Kepler finds, in fact, four "skeletons" or frameworks (*sceleta*) for six-note chords, two of them durus and two mollis, that include at least one possible motion from each of the planets. The characteristics of these skeletons are that all the notes harmonize together in a polyphonic chord and that each note lies within the range of some planet's apparent motion. When two or more notes within a planet's range can harmonize with the other notes, each is included in the skeleton as a possibility. A single skeleton thus represents a class of possible chords, only one of which can occur at any one moment.

A skeleton is also characterized by its latitude of tension (*tensionis latitudo*), or range of possible tuning. Since the apparent speed of each planet varies continually, any single chord can be played throughout a range of pitches. So long as all the notes in a chord remain within the limits of their respective planets' possible motions, the entire chord can be raised or lowered in pitch and still be attainable. Increasing or decreasing the assigned speeds of each planet proportionally does not affect the ratios among those speeds and hence leaves the chord intact. The chord's latitude of tension is the overall range within which this can be done. The lower end of the range is the tuning at which the note assigned to any one planet reaches the planet's aphelial motion, for if the tuning of the chord is lowered beyond that point the planet necessarily falls silent and the chord is no longer universal. The upper end of the range is similarly the tuning at which some planet's note reaches the perihelial limit of its motion, for if the chord is tuned any higher that planet will no longer join in. It is obvious that there is an upper limit on the latitude of tension: no chord can be tuned over a range greater than the proper proportion of Venus's motions without losing the participation of that planet.

When a skeleton represents more than one chord, the skeleton itself has a range of tuning that reaches as low as the range of any of its chords, and as high as the range of any of its chords. When a planet has several possible notes in a skeleton, raising the overall pitch silences them one at a time, as each reaches the highest allowable pitch for that planet at its perihelial motion. So long as any one of the planet's notes remains, however, the skeleton is still a valid framework for at least one six-note chord.

It is not as difficult as it might seem to find all possible skeletons of universal harmony. Venus and Earth, the planets with the smallest eccentricities, must be included in any six-note chord. The only harmonious intervals possible between these two planets are the major and minor sixths, and as we shall see, even those intervals are possible only through quite a limited range of tuning. Working outward from this small range, Kepler determines in chapter 7 which possible motions of the other planets fit into a chord with the major or minor sixth between Venus and Earth.

Such chords are not easily written. To begin with, they are gigantic. Since Mercury's apparent perihelial motion is more than two hundred times Saturn's apparent aphelial motion, the universal chords span well over seven octaves—a range greater than that of a modern piano. In writing down the chords, the actual note assigned to any particular motion is arbitrary. Kepler begins his chords at G on the bottom line of the bass clef and simply continues to draw staves above it, so that the top notes are more than twenty ledger lines above the staff.

THE DURUS SKELETONS

Figure 9.10 shows the first two skeletons, containing the possible durus chords of all six planets.[47] For the wide range of notes in this table and the next, I use Kepler's notation of octaves. Successive octaves on A are indicated by the sequence A, a, aa, aaa, aaaa, and so forth. The style changes after the note A in each octave: G, A, b, c, . . . , a, bb, cc, and so on.[48] For these enormous chords Kepler condenses the notation further, indicating octaves with a small Roman numeral. Thus b^i is bb, and e^{vii} is eeeeeee, seven octaves above simple e. Mercury always appears on multiple notes within each skeleton, and Mars and Saturn sometimes do so. Despite the appearance of the

---

[47] As noted above, I use B♭ for Kepler's B, and B for his H. Kepler himself wrote E♭ on musical staves, but D♭ (for D♯) in text, as here in the labels.

[48] *G.W.* 6.148.21–26.

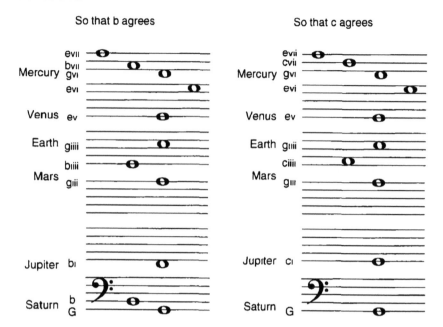

FIGURE 9.10. Universal harmonies of the durus type

staves, no lines are omitted between them (in the way the line for middle C is omitted between the staves of the bass and treble clefs). Middle C, sounded by Jupiter on the right side of the figure, lies on the bottom line of the second staff from the bottom, where it is denoted c¹.

Table 9.8 reproduces the numerical values and notes accompanying these two skeletons of the possible universal harmonies of the durus type. The skeleton on the left includes four possibilities for Mercury, two for Mars, and two for Saturn. Since each possible note for each planet creates a distinct six-note chord, the skeleton represents 4 × 2 × 2, or 16, different chords. The skeleton on the right, with four possible notes for Mercury and two for Mars, gives 4 × 2, or 8, possible universal chords.

These are all durus chords with G in the bass, containing the notes B or E rather than B♭ or E♭. Kepler labels them according to whether B or C is present in the chord. In modern terminology the chords in the skeleton on the left are E-minor chords, while those in the skeleton on the right are C-major chords. The columns of lowest and highest tunings show the limits of the latitude of tension. Absent notes in one of these columns, such as Mercury's eᵛⁱⁱ in the highest tuning, lie outside the planet's possible range of speed. That eᵛⁱⁱ would correspond in the

TABLE 9.8.
Universal Durus Harmonies

1		2	3	4		5	6
		So That B Agrees				*So That C Agrees*	
	tuning:	*lowest*	*highest*		*tuning:*	*lowest*	*highest*
Mercury	e^{vii}	380'20"	none	Mercury	e^{vi}	380'20"	none
	b^{vi}	285'15"	292'48"		c^{vii}	304'16"	312'21"
	g^{vi}	228'12"	234'16"		g^{vi}	228'12"	234'16"
	e^{vi}	190'10"	195'14"		e^{vi}	190'10"	195'14"
Venus	e^{v}	95'5"	97'37"	Venus	e^{v}	95'5"	97'37"
Earth	g^{iiii}	57'3"	58'34"	Earth	g^{iiii}	57'3"	58'34"
Mars	b^{iiii}	35'39"	36'36"	Mars	c^{iiii}	38'2"	39'3"
	g^{iii}	28'32"	29'17"		g^{iii}	28'32"	29'17"
Jupiter	b^{i}	none	4'34"	Jupiter	c^{i}	4'45"	4'53"
Saturn	b	2'14"	none				
	G	1'47"	1'49"	Saturn	G	1'47"	1'49"

In this universal harmony agree Saturn with its aphelial motion, Earth with its aphelial, Venus with almost its aphelial; in the highest tuning Venus agrees with its perihelial; at an intermediate tuning agree Saturn with its perihelial motion, Jupiter with its aphelial, Mercury with its perihelial. So Saturn can agree with two motions, Mars with two, Mercury with four.

All are still here except the perihelial motion of Saturn and the aphelial of Jupiter; but Mars with its perihelial motion agrees in their place. They agree with single motions other than Mars, with two, and Mercury, with four.

highest tuning to a motion of 390'28" (twice the 195'14" of e^{vi}) and hence would be faster than Mercury's perihelial motion of 384'. (Extreme motions cited in this and subsequent paragraphs can be found in table 9.3.)

It makes little difference if one of a planet's multiple notes is forced to drop out in this way, but if a planet's only note disappears, the harmony ceases to be universal beyond that point. And indeed, this has happened to Jupiter, which is entirely missing from the lowest tuning on the left side. In that tuning Jupiter's b^{I} would correspond to a motion of 4'27",[49] which is slower than Jupiter's aphelial motion. Consequently Kepler has not entered a note for Jupiter under the lowest tuning, and that tuning does not truly yield a universal harmony. Given Jupiter's aphelial motion of 4'30", the lowest tuning yielding a six-note chord would show Saturn's G and b rounded to 1'47" and 2'15", Jupiter's b^{I} at 4'30", and so forth.

A similar mishap has befallen Mars, whose c^{iv} (at the right) just barely exceeds that planet's perihelial motion of 38'1" in both tunings. So small a difference would not be "audible." Luckily Mars also plays g^{iii} in that skeleton, so four universal chords remain in the highest tuning of the skeleton, even if one denies the other four that assign Mars to c^{iv}.

Kepler neglects these details because he is concentrating on Venus and Earth, the two planets that impose the tightest constraints on the universal harmony. The motion of Venus varies only by about 34:35; but the latitude of tension must be even less than this in durus harmony if Earth is to play in the same chord. The only harmonic intervals possible between Earth and Venus are the major and minor sixths. In the durus scale the sixth has to be a major sixth, corresponding to a proportion of 3:5. The limits of the skeleton's tuning are determined by this major sixth, as follows.

The perihelial motion of Venus is 97'37". A major sixth below this is 58'34", which is within the observed range for Earth. Thus the perihelial motion of Venus fixes the highest tuning in both durus scales, for at higher tunings Venus must drop out of the chord. The aphelial motion of Venus, however, is too slow to define the lowest tuning. If one were to lower the pitch of the entire skeleton by 34:35 to use the aphelial motion of Venus, Earth's note g^{iiii} would be forced down to 56'54", which is impossible because it is less than Earth's aphelial motion. At so low a tuning no durus chord can include both Earth and Venus. Hence it is the aphelial motion of Earth that constrains the lowest tuning of the chord "containing B."

One speculates that Kepler failed to notice until the book was in press that even this tuning still forces the note b^{i} out of Jupiter's range. The absence of a planet from one of the columns in a table entitled

---

[49] The lowest tuning is defined by the perihelial motions of Earth, 57'3", and of Venus, 95'5". Four octaves and a fourth below Venus's e^{v} is 95'5" ÷ 16 ÷ $\frac{4}{3}$ = 4'27"; likewise three octaves and a minor sixth below Earth's g^{iv} is 57'3" ÷ 8 ÷ $\frac{8}{5}$ = 4'27".

"Harmonies of All the Planets" certainly suggests a very late correction; if type had not been set, Kepler could easily have recalculated the lowest tuning to include Jupiter. As it is, he has deleted the out-of-range pitch for Jupiter but retained the lowest tuning even though it is no longer truly universal.

Chords containing as many as possible of the extreme motions are naturally the rarest and most excellent of the universal harmonies. Kepler includes the c^{iv} of Mars in his chords, although in both tunings it is slightly higher than the Martian perihelial motion allows, because he wants to see as many of the extreme motions as possible in the universal harmony, and c^{iv} is very close to the perihelial motion of Mars. In his notes beneath the two skeletons he carefully points out that the first of them accommodates the aphelial motions of four planets, and the perihelial motions of three planets, at some point within its latitude of tension. The second skeleton, the one to which C agrees, includes three aphelial and three perihelial motions.

THE MOLLIS SKELETONS

Kepler gives two further skeletons showing the possible mollis chords that include all six planets (figure 9.11). The staves in this figure appear to be unnecessarily festooned with flats, many of them not applied to any note in particular. These are characteristic of Kepler's musical notation. He regards the ♭ symbol as the letter b, which is how one writes the note b♭ in German. In book 3 he has explained that the presence of a ♭ on the staff is properly an indication of where the note b♭ is located, although over time it has come by analogy to indicate other lowered notes such as E♭—but only on a musical staff.[50] (Kepler never uses the ♭ symbol in text but always writes D♯ for E♭, for example, and G♯ for A♭.) Hence most of the ♭ symbols in figure 9.11 are not so much flat symbols as B-flat symbols. As such they are actually rather handy, serving to point out the location(s) of the note B♭ on each staff.

As with the durus skeletons, I have extracted the notes and the corresponding motions into a table (table 9.9). This time Kepler is able to use the full range of the extreme motions of Venus to determine the skeletons' ranges of tuning, since Earth can stay a minor sixth below any possible motion of Venus. There are sixteen chords in the left half of the table, including any of four notes for Mercury, either of two for

[50] *G.W.* 6.147.29–34.

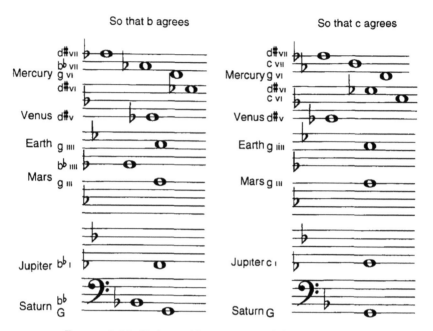

FIGURE 9.11. Universal harmonies of the mollis type

Mars, and either of two for Saturn. On the right side of the table are five chords, identical except for the note assigned to Mercury.

These chords contain the mollis thirds and sixths from G, namely B♭ and D♯. We would say that the skeleton on the left contains E♭-major chords and that the one on the right contains C-minor chords. Kepler calls the left skeleton mollis chords "to which h agrees," in his German notation, evidently intending b♭—b for him—as the kind of h that belongs in a mollis chord. The left column of his table is headed *Ut h concordet,* but the notes in it include b, not h. (On the other hand, perhaps the printer inadvertently copied column headings from the previous table.)

There are no serious problems in these chords. As before, the range of tuning is slightly too large for some of the possible notes at the extreme tunings. Mercury's c^{vi} is a bit below the aphelial motion of that planet. It is included for the same reason that an out-of-range note for Mars appears in the earlier durus chords: Kepler wants as many as possible of the extreme motions in the skeletons, and at the highest tuning of this skeleton c^{vi} does come quite close to Mercury's aphelial motion. The exact motions corresponding to particular notes in the mollis skeleton have all shifted a bit from the durus skeleton—notice

TABLE 9.9.
Universal Mollis Harmonies

1	2	3	4	5	6
	So That B[b] Agrees			So That C Agrees	
tuning:	lowest	highest	tuning:	lowest	highest
Mercury d#vii	379'20"	none	Mercury d#vii	379"20'	none
bb vii	284'32"	292'56"	cvii	316'5"	325'26"
gvi	237'4"	244'4"	gvi	237'4"	244'4"
d#vi	189'40"	195'14"	d#vi	189'40"	195'14"
			cvi	none	162'43"
Venus d#v	94'50"	97'37"	Venus d#v	94'50"	97'37"
Earth giiii	59'16"	61'1"	Earth giiii	59'16"	61'1"
Mars bbiiii	35'35"	36'37"			
giii	29'38"	30'31"	Mars giii	29'38"	30'31"
Jupiter bbi	none	4'35"	Jupiter ci	4'56"	5'5"
Saturn bb	2'13"	none			
G	1'51"	1'55"	Saturn G	1'51"	1'55"

Here again, at an intermediate tuning agree Saturn with its perihelial motion, Jupiter with its aphelial, Mercury with its perihelial. At the highest tuning the perihelial motion of Earth nearly agrees.

And here, with the aphelial motion of Jupiter and the perihelial of Saturn excluded, the aphelial motion of Mercury is very nearly admitted in addition to the perihelial. The others remain.

that Saturn's G is faster—because Venus, whose scarcely varying motion sets the pitch, is now singing d#v instead of ev.

Once again, Kepler has appended a detailed note below each side of his table to point out how many of the extreme motions appear in the skeletons. The mollis chord to which Bb agrees includes one aphelial and three perihelial motions. The chord to which C agrees includes one near-aphelial motion, Mercury's just-out-of-range cvi, and the same three perihelial motions.

Kepler concludes from his skeletons of universal harmony that

Astronomical experience has revealed that universal harmonies of all the motions can occur, and they are of two types, durus and mollis, and of two forms or, if I may say so, modes; and in each one of the four cases,

179

with a certain latitude of tension, and also with a certain variety of the particular harmonies of Saturn, Mars, and Mercury, and of each with the others. Nor does this occur only with the intermediate motions, but with the extreme motions of all, aside from the aphelial of Mars and the perihelial of Jupiter.

This last point recurs; it is important to Kepler. The harmonies do not depend upon the infinite variety of intermediate motions but are dignified by the presence of the aphelial and perihelial motions. In fact, all the extreme motions except two are found in these four great chord-skeletons. The intractable pair of extreme motions are the aphelial motion of Mars and the perihelial motion of Jupiter. Both these motions are too close to being an exact number of octaves below Venus's monotone ever to harmonize with it. Kepler justified the right of Venus to shut out these last two extreme motions in a single long and astonishing sentence, building an elaborate metaphor of Earth and Venus as husband and wife:

> This [i.e., the admission of these last two motions into the universal harmony] has, as an impediment, the marriage between Earth and Venus—the two planets that distinguish the types of harmony into the hard and masculine, and the soft and feminine—as husband and wife: depending on which spouse has indulged the other, either Earth is in his aphelion, maintaining the husbandly dignity and doing a man's work, with Venus in her perihelion, as if withdrawn and sent off to take care of women's business; or else she has received him, climbing seductively into her aphelion, or Earth has plunged into his perihelion close to Venus, and into her embrace, seeking pleasure with shield put aside and weapons and man's work: for that is when the harmony is soft.[51]

OMITTING THE OBSTRUCTIVE COUPLE

Kepler returns to his main argument to see what will be possible "if we bid this obstructive Venus to be quiet." He constructs chord skeletons, in other words, from the motions of the other five planets. His goal in doing this is simply to include as many as possible of the extreme motions in the chords. By leaving out the E sung by Venus, he is able to admit the previously disallowed perihelial motion of Jupiter at D. This D enters into durus and mollis chords that we recognize as G major

---

[51] Ibid., 6.326.31–327.9. Hard and soft are of course the literal meanings of durus and mollis.

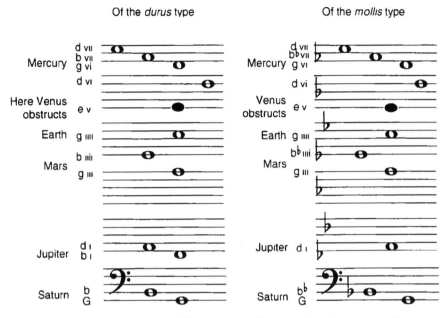

FIGURE 9.12. Harmonies of five planets, disregarding obstructive Venus

and G minor.[52] Figure 9.12 displays the possible five-note chords in these skeletons, and table 9.10 shows the motions involved in the chords.

Below the durus skeleton Kepler notes that it includes the aphelial motions of Saturn and Earth, in its lowest tuning; the perihelial motion of Saturn and the aphelial of Jupiter, in an intermediate tuning; and the perihelial motion of Jupiter in its highest tuning. The mollis skeleton does not allow Jupiter's aphelial motion but, he claims, comes close in its highest tuning to the perihelial motion of Saturn. This at first seems odd, since those two motions are just an octave apart. Saturn's perihelial bʹ in the highest tuning is, however, slightly *inside* the range of that planet's motions, falling short of the extreme motion by half a diesis; whereas Jupiter's aphelial bʹ would be *outside* the planet's range of motion by that half-diesis.

---

[52] Ibid., 6.327. The latitude of tuning in these chords is constrained by the aphelial motion of Earth singing G at the low end of the range, and the perihelial motion of Jupiter singing D at the high end. Kepler carelessly omits the motion 0;1,50, corresponding to Saturn's G in the highest tuning, from the table for the durus chord of five planets. In *G.W.* this table (on 6.327) also has a misprint on the mollis side for the highest tuning of obstructive Venus; this motion should be 97,37 as on the durus side.

TABLE 9.10
Harmonies of Five Planets, Disregarding Venus

1		2	3	4		5	6
	Of the Durus type				Of the Mollis type		
	tuning:	lowest	highest		tuning:	lowest	highest
Mercury	d^{vii}	342'18"	351'24"	Mercury	d^{vii}	342'18"	351'24"
	b^{vii}	285'15"	292'48"		$b\flat^{vii}$	273'50"	280'57"
	g^{vi}	228'12"	234'16"		g^{vi}	228'12"	234'16"
	d^{vi}	171'9"	175'42"		d^{vi}	171'9"	175'42"
Venus obstructs here	$[e^{v}]$	95'5"	97'37"	Venus obstructs	$[e^{v}]$	95'5"	97'37"
Earth	g^{iiii}	57'3"	58'34"	Earth	g^{iiii}	57'3"	58'34"
Mars	b^{iiii}	35'39"	36'36"	Mars	$b\flat^{iiii}$	34'14"	35'8"
	g^{iii}	28'31"	29'17"		g^{iii}	28'31"	29'17"
Jupiter	d^{i}	5'21"	5'30"	Jupiter	d^{i}	5'21"	5'30"
	b^{i}	none	4'35"				
Saturn	b	2'13"	none	Saturn	b♭	2'8"	2'12"
	G	1'47"	[1'50"]		G	1'47"	1'50"

Here Saturn and Earth agree in the lowest tuning with their aphelial motions; in an intermediate tuning Saturn with its perihelial, Jupiter with its aphelial; in the highest tuning, Jupiter with its perihelial motion.	Here the aphelial motion of Jupiter is not allowed, but in the highest tuning Saturn agrees with a motion close to its perihelial.

Continuing his pursuit of the planets' extreme motions, Kepler next asks the husband of obstructive Venus to stop singing g^{iiii}. This enables him to obtain the four-planet chord skeletons in figure 9.13.[53] Saturn's bass note has moved from G to b, on the left, and to A, on the right. The chords on the left are mollis, since they contain d, which is a minor

[53] In Kepler's first edition of 1619, and also in Frisch's *Opera Omnia* and Caspar's *G.W.*, the note b^{iiii} on the left side of this figure is misplaced. It appears on the second line of the fourth staff from the bottom, but it should be on the bottom line of that staff as shown here.

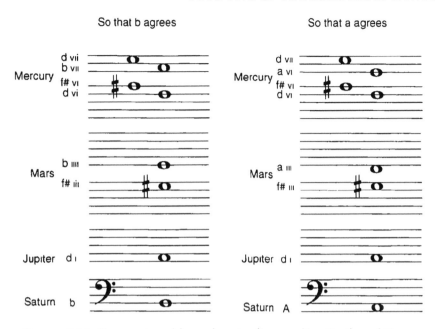

FIGURE 9.13. Harmonies of four planets, disregarding Earth and Venus

third above b; and the chords on the right are durus, since they contain f♯, which is a major sixth above the A at the bottom. In modern terminology the mollis chord on the left side is B minor and the durus chord on the right side is D major.

The whole point of dropping Earth from the harmony is to allow the last unharmonized extreme motion, the aphelial motion of Mars, to enter the chords. This is now possible. Since Saturn's G has moved to A or b, the f♯iii of Mars's aphelial motion is no longer dissonant. The motions involved in these chords are shown in table 9.11. Kepler omits the numbers for the motions in the chord to which A agrees; I have supplied them in column 6 of the table.

Kepler says that these four-planet harmonies are without latitude of tension, but I do not understand why. The aphelial motion of Mars prevents him from assigning slower motions to the notes, to be sure; but he could give them slightly faster motions until Saturn's b, in the mollis skeleton at the left, ascends from 2'11" to its perihelial limit of 2'15". The durus chords can go higher yet, until Jupiter's d¹ reaches its perihelial limit of 5'30". At any rate, Kepler provides motions for only a single level of tuning in the mollis skeleton and omits them altogether in the durus skeleton.

TABLE 9.11.
Harmonies of Four Planets, Disregarding Venus and Earth

1	2	3	4	5	6
So That B Agrees [Mollis]			So That A Agrees [Durus]		
Mercury	d^{vii}	335'50"	Mercury	d^{vii}	[335'50"]
	b^{vii}	279'52"		a^{vi}	[251'52"]
	f♯^{vi}	209'52"		f♯^{vi}	[209'52"]
	d^{vi}	167'55"		d^{vi}	[167'55"]
Mars	b^{iiii}	34"59"	Mars	a^{iii}	[31'29"]
	f♯^{iii}	26'14"		f♯^{iii}	[26'14"]
Jupiter	d^{i}	5'15"	Jupiter	d^{i}	[5'15"]
Saturn	b	2'11"	Saturn	A	[1'58"]

Kepler has found a place for all twelve of the extreme motions in eight great chordal skeletons—not, alas, all at once. The perihelial motion of Jupiter was dissonant until he dismissed Venus from the choir; and in order to harmonize the aphelial motion of Mars he was forced to excuse Earth. Venus and Earth, the two planets with the most nearly circular orbits, blocked any straightforward realization of universal harmony in the heavens. Unless he finds a reason for the obstructions, Kepler cannot easily claim to have explained the structure of the world.

He is unabashed. The heavens themselves are singing polyphony. For Kepler, "it is no longer surprising that the theory of singing in polyphonic harmony, unknown to the ancients, has at last been found by Man, the Ape of his Creator."[54]

CHAPTER 8: THE FOUR VOICES OF CELESTIAL HARMONY

In this brief chapter Kepler proposes that the division of a human chorus into soprano (discantus), alto, tenor, and bass parts is also prefigured in the heavens. He concedes that heavenly harmonies involve no voices or sounds, "because of the great tranquility of the motions," and that there is no reason in the heavens why four parts

[54] G.W. 6.328.25–27.

should be used rather than some other number. Indeed, there are six planets—this is certain, because there are five regular polyhedra—but his admiration for the analogy between human song and heavenly motion compels him to remark on the analogy between the roles of the planets and of those of the singers in a choir.

In book 3 he characterized the four voices. The bass voice is deep and slow, jumping across harmonic intervals. The tenor and alto, intermediate voices, have limited range, particularly the alto. The soprano voice is high-pitched and rapid, moving quickly by small steps.[55] Here in book 5 he points out that Saturn and Jupiter exhibit the properties of the bass, Mars of the tenor, Earth and Venus of the alto, and Mercury of the soprano. He gives five reasons supporting this analogy.

First, in song the bass voice is set against (*opponitur*) the alto, and in his analogy only these voices are associated with two planets each. Second, the alto voice is nearly the highest and is confined within a narrow range—both traits suggestive of Venus and Earth. Third, the tenor is "free, but proceeds modestly," which reminds Kepler of his old enemy Mars, whose proper interval of a fifth is exceeded only by Mercury. Fourth, the bass moves by harmonic leaps such as those among the various extreme motions of Saturn and Jupiter. Finally, the soprano voice has the largest range and moves the fastest, in obvious analogy to Mercury.

"This might all be accidental," Kepler acknowledges. "Let us listen now to the causes of the eccentricities."[56]

CHAPTER 9: THE CAUSES OF THE ECCENTRICITIES

In chapter 9 of book 5 Kepler consummates his long-avowed intention of exhibiting the Creator's detailed design for the heavens. It is a very long chapter, thirty folio pages out of book 5's seventy-two in the first edition. Max Caspar has accurately described the experience of reading it:

> Anyone who follows [Kepler] step by step soon finds himself lost in a thicket of cross-references, so that he no longer knows where the tortuous path leads and, astonished, asks himself how his guide could

---

[55] Ibid., 6.181.1–19.
[56] Ibid., 6.329.

comprehend all this. It is in all ways the most peculiar thing that one can imagine. He who wrote this chapter must for many days have completely forgotten everything around him, and lived in an extreme concentration of the spirit, as in an ecstasy. Otherwise it would not have been possible to comprehend the links of this chain of thought, to bring the last ones into agreement with the first, and to reconcile the entangled numerical proportions to the smallest detail.[57]

If one pays diligent attention while following Kepler through his thicket, one finds that the entangled numerical proportions are here and there not reconciled as exactly as this passage implies. It is nonetheless patently true that the author existed *in einer höchsten Konzentration des Geistes* when he composed chapter 9.

The chapter is entitled in full "The Origin of Eccentricities in Individual Planets from Attention to the Harmonies among Their Motions." In its final form it consists of forty-eight numbered sections, labeled either as axioms or as propositions, with proofs—varying widely in character—for most of the propositions. A forty-ninth section, labeled *Epiphonema*, or summary, affirms how good it is that the different kinds of reasons adduced in the first forty-eight sections should have taken precedence in the order that they have been found to take precedence.

The propositions, it is perhaps worth noting, are not stated with infinitive verb forms, which mathematical theorems in Latin typically use, nor in a present tense, which Kepler used in stating similar series of propositions in books 1, 2, and 3, and in chapter 3 of book 5. These in chapter 9 are almost all in a past tense, because they are about choices the Creator made while laying out the primordial structure of the world. Those earlier propositions were abstract; these are concrete and even, in a sense, historical.

Prior and Posterior Reasons

Kepler divides the forty-nine sections of chapter 9 into two unequal parts, the *rationes priores* of the first seventeen sections and the *posteriores rationes* making up the last thirty-two. This distinction is not, as it might appear, between a priori and a posteriori reasons but rather between two groups of reasons worked out with different themes and at different levels of detail.

---

[57] M. Caspar, in his *Nachbericht* to the *Harmonice mundi*. (*G.W.* 6.500).

The themes of the *rationes priores* include a broad principle of variety in the universe; the role of the regular polyhedra in establishing the basic structure; and the relative dignities of the planets and of the polyhedra. Such considerations, which had occupied Kepler since the 1590s, suffice to explain most—but not all—of the proportions among extreme apparent motions. The propositions cannot by any means be characterized as "easy." Starting with proposition 6 they depend upon the sophisticated planetary theory that Kepler summarized at the end of chapter 3; and they cannot have been written out until after the discovery of the period-distance relation in May of 1618.

In order to account for the last and most dissonant of the heavenly proportions, those involving "obstructive Venus" and the Earth, her husband, Kepler makes a new start with the *posteriores rationes*. These later arguments are based on careful analysis of the roles played in the polyphonic harmony of the heavens by the two most problematic planets. He distinguishes durus and mollis harmonies and identifies them as essential components in the cosmic structure. The propositions become more difficult, the logic relating them more intricate, and the difficulties more obscure.

All this is to resolve some relatively small problems. As Kepler admits toward the end, the second group of arguments modifies some results of the first group, but in details only, "which in Harmony means in the intervals smaller than the *concinna.*"[58]

Early Axioms: That Variety Might Adorn the World

Kepler characterizes nine of the numbered sections as axioms. By this he means not that their truth is "self-evident," and certainly not that they are free postulates, but rather that they are the basic principles on which his reasoning is founded. He undoubtedly believes them to be true. They express two themes: the role of the regular polyhedra, and a principle of variety or abundance that Kepler finds embodied in Creation.

The harmonic theorems of chapter 9 are explicitly founded upon the regular polyhedra, whose adequacy in explaining the rough proportions of the planetary system Kepler has established more than once: in the opening chapters of book 5; in book 4 of his *Epitome astronomiae Copernicanae*; and indeed two decades earlier in the

---

[58] *G.W.* 6.345.13–15.

Mysterium cosmographicum. Toward the end of this ninth chapter, Kepler will insert a further polyhedral "axiom" asserting that the solid figures would have determined the proportions of the orbits exactly, had they not been prevented from doing so by the intricate harmonic considerations he has discussed in the meantime.

Most of the other axioms require in one way or another that Creation should contain as many good things as possible. This assumption, to Kepler, is not controversial. He establishes it immediately as axiom 1.

1. *Axiom.* It is appropriate, everywhere that it is possible, that all types of harmonies should have been established between the extremes of the motions both of individual planets and of pairs, so that this variety might adorn the world.

The second axiom announces the regular polyhedra as the fundamental model for the celestial world:

2. *Axiom.* The five intervals between the six planets needed to correspond in size, to a certain extent, to the proportion of the geometric spheres that are inscribed and circumscribed to the five regular solid figures; and in the same order that is natural to the figures themselves.

Concerning this, see chapter 1 and the Mysterium Cosmographicum *and book 4 of the* Epitome of Astronomy.

The next axiom, numbered 4 after an intervening proposition, returns to the principle of variety:

4. *Axiom.* The planets all need to have different eccentricities and different motions in latitude; and consequently different distances from the Sun, the source of motion.

By varying its altitude from the Sun and its latitude from the ecliptic, he explains, each planet weaves a three-dimensional structure, like a cocoon.[59] Changing solar distances lead, by Kepler's physical principles, to changing rates of motion, since a planet moves more slowly when it is more distant from the source of its motion. This diversity of motions multiplies the possible harmonies among them.

The last of the early axioms is a handy tiebreaker:

10. *Axiom.* When other things are equal, the proper proportion that is more important, more excellent, or simply larger, is due to the higher planet.

[59] Ibid., 6.331.14–16; 332.2–12.

Prior Reasons: Harmonic Implications of the Polyhedra

On the foundation of his polyhedral hypothesis, Kepler sets out immediately to establish the harmonies of the planetary system from first principles. Thus:

> 3. *Proposition.* The intervals between Earth and Mars, and between Earth and Venus, had to be smallest in proportion to their spheres, and about equal to one another. Those between Saturn and Jupiter, and between Venus and Mercury, had to be intermediate, and again about equal. That between Jupiter and Mars had to be greatest.

The proportional distance between the first two pairs of planetary spheres, those separated by the dodecahedron and icosahedron, is just the ratio of the inscribing to circumscribing spheres for those figures. He knows from geometry that this ratio is smallest (and equal) for these two figures. Similarly the ratios between inscribing and circumscribing spheres for the other figures indicates the proportional distances between the other pairs of planetary spheres, which must consequently be in the stated order.

> 5. *Proposition.* To each pair of neighboring planets, two different harmonies had to be assigned.

These are the proportions of the converging motions, the perihelial motion of the superior planet with the aphelial of the inferior; and of the diverging motions, the aphelial of the superior with the perihelial of the inferior. The two proportions for each pair of planets are different because each planet's orbit is eccentric, rather than circular, by axiom 4. Proportions between different pairs of planets are different so that this variety too might adorn the world.

The polyhedra rule out some harmonies altogether:

> 6. *Proposition.* The two smallest harmonic proportions 4:5 and 5:6 do not occur between any pair of planets.
>
> 7. *Proposition.* The harmonic proportion of a fourth cannot occur between the converging motions of two planets, unless for them the combined proper proportions of extreme motions are more than a fifth.

In proving proposition 7, Kepler considers first the simple case of circular orbits, where both motion and distance are constant. He supposes that the proportion of apparent, or angular, motions is 3:4. The proportion of periodic times is therefore 3:4 inversely. The period-distance relation required that the proportion of distances be $\frac{2}{3}$ of 3:4, or about 825:1,000 (we would calculate $(\frac{3}{4})^{\frac{2}{3}} \cong \frac{825}{1,000}$). None of the regular polyhedra allows its two spheres so close as that.

A converging proportion of 3:4 is possible between two *noncircular* orbits of sufficient eccentricity, because on an eccentric orbit a planet's apparent velocity varies more than does its distance. For an ellipse of given semimajor axis and variable eccentricity, the apparent velocity at the apsides varies inversely as the square of distance. The polyhedra prevent neighboring planets from approaching each other too closely in actual distance, regardless of eccentricity; but their apparent velocities can approach each other more closely on eccentric orbits than on circular orbits. Kepler calculates the necessary eccentricities, with the aid of some rather obscure intermediate results from his proof of the thirteenth proposition from chapter 3 of book 5.

Suppose that the proportion of the converging motions of two neighboring planets is a fourth, 3:4. Using the notation m, M for the apparent motions of the inner and outer planets, and r, R for their distances from the Sun, the supposition is that:
$$M_p:m_a = 3:4 = 750:1,000$$
Suppose also that the planets are as close as allowed by the regular polyhedra. The ratio of spheres for both the dodecahedron and the icosahedron is 795:1,000; hence
$$r_a:R_p = 795:1,000$$
The excess (*excessus*) of the proportion of motions over that of distances is thus:
$$M_p:m_a - r_a:R_p = 750:1,000 - 795:1,000 = 750:795$$
Referring back to equation (9) from our analysis of chapter 3, we see that this *excessus* is precisely the term d from that demonstration.

Equation (8) in our analysis of chapter 3 stated[60] that
$$\tfrac{1}{2}r:R + d = r_a:R_p \tag{8}$$
Here Kepler claims that subtracting the *excessus* from the proportion $r_a:R_p$, "according to the theory of chapter 3," leaves half the proportion of the spheres:
$$\tfrac{1}{2}r:R = r_a:R_p - d$$
This obviously follows immediately from (8). Using the above num-

---

[60] Ibid., 6.305.3–4.

bers, taken from the assumption that the proportion of converging motions is 3:4, he calculates:

$$\tfrac{1}{2}r{:}R = 795{:}1{,}000 - 750{:}795$$
$$= 7{,}950{:}9{,}434, \text{ approximately.}$$
$$\text{So } r{:}R = 7{,}950{:}9{,}434 + 7{,}950{:}9{,}434$$
$$= 7{,}100{:}10{,}000, \text{ approximately.}$$

But the proportion $r{:}R$ has three components.

$$r{:}R = r{:}r_a + r_a{:}R_p + R_p{:}R$$
$$r{:}R - r_a{:}R_p = r{:}r_a + R_p{:}R$$
$$7{,}100{:}10{,}000 - 795{:}1{,}000 = r{:}r_a + R_p{:}R$$
$$r{:}r_a + R_p{:}R = 7{,}100{:}7{,}950$$

The sum of the two proportions between the mean and converging distances would equal this interval, 7,100:7,950—about 8:9, or a major tone. The sum of the proportions between diverging and converging distances would be about twice this, two major tones. The sum of the proportions between diverging and converging *motions*—that is, the sum of the proper proportions of the planets' motions—would be about twice that of the proportions of distances, or four major tones. This interval, Kepler points out, is more than a fifth; actually it is more than a minor sixth. Certainly if the sum of the planets' proper proportions of motions were less than a fifth, the proportion of their converging motions could not be as small as a fourth. Otherwise their orbits would be too close for any of the regular polyhedra to fit in between.

An analogous conclusion follows for the major and minor thirds ruled out in proposition 6. Using the major third 4:5 instead of the fourth 3:4 in the above equations gives

$$d = M_p{:}m_a - r_a{:}R_p$$
$$= 4{:}5 - 795{:}1{,}000$$
$$= 800{:}795$$

The *excessus* is now a *defectus*, since the proportion of converging distances is greater than the hypothesized proportion of converging motions. Working through the equations,

$$\tfrac{1}{2}r{:}R \quad = 795{:}1{,}000 - 800{:}795$$
$$= 795{:}1{,}000 + 795{:}800$$
$$\cong 790{:}1{,}000$$
$$\text{So } r{:}R \cong 624{:}1{,}000$$

Separating $r{:}R$ into its components, as above, and subtracting the proportion of converging distances from this last equation, leaves the sum of the proportions of mean to converging distances for this hypothesis (which, if the reader has forgotten, is that the proportion of the converging motions of a pair of planets is a major third). Numerically,

$$r:r_a + R_p:R = 624:1,000 - 795:1,000$$
$$= 624:795$$

This is more than a major third. As before, the sum of the proper proportions of extreme motions is about four times as much. It amounts to 380:1,000, nearly an octave and a fourth. This exceeds the sum of the proper proportions observed for any pair of adjacent planets.

One concludes that it is impossible for the proportion of the converging motions of two adjacent planets to be as small as a major third, unless the sum of the proper proportions of their extreme motions is more than an octave and a fourth: huge, as Kepler says,[61] without explicitly working out this argument for the major third. (We may note that Mercury by itself has a proper proportion of an octave and a major third. Its neighbor Venus, however, is so little eccentric that the pair still fails the criterion. Besides, an even greater interval would be required between the orbits of those two planets to accommodate the octahedron that belongs there.)

Propositions 6 and 7 are the most sophisticated of the *rationes priores* that form the first part of chapter 9. The major and minor thirds are absolutely blocked, as converging intervals, by the regular solid figures; and the fourth is allowed only if the sum of the two planets' proper proportions is greater than a fifth. The demonstrations of these propositions have depended upon the advanced planetary theory from the end of chapter 3, including the period-distance relation that Kepler discovered in May of 1618. Most of the remaining propositions in the *rationes priores* depend upon one of these two propositions, or directly upon the thirteenth section of chapter 3. Even these early propositions in chapter 9, therefore, were written down during or after May of 1618.

Turning to the task of assigning specific proportions to specific planets, Kepler sets out a series of propositions, based largely on archetypal reasoning, wherein he establishes most of the planetary harmonies with disconcerting ease.

8. *Proposition.* To Saturn and Jupiter were due the harmonies 1:2 and 1:3, that is, the octave and the octave-plus-fifth.

---

[61] Ibid., 6.332.34–35.

Saturn and Jupiter, as the first and highest of the planets, deserve the "first" harmonies, 1:2 and 1:3.[62] This is not merely a presumption, for it is supported by mathematical analysis. The cube between the spheres of these two planets implies that the various ratios of their distances are all about $1:\sqrt{3}$, the cube's proportion of spheres. Using results from chapter 3 in enormously simplified form,[63] Kepler infers that the converging proportion of Saturn and Jupiter must be the next-greater harmonic proportion to this, namely 1:2. The harmony 1:3, an octave and a fifth, remains for the diverging proportion of Saturn and Jupiter.

Other arguments support the same conclusion. The proportion of diverging motions is much greater than the sesquialterate of the proportion of spheres, as shown in chapter 3, and thus must be close to twice that proportion. The cube's proportion of spheres is $1:\sqrt{3}$, so again the proportion 1:3 is suitable for the proportion of Saturn's and Jupiter's diverging motions.

   9. *Proposition.* The combined proper proportions of the extreme motions of Saturn and Jupiter had to be about 2:3, a fifth.

The proportion of these planets' diverging motions is composed of Saturn's proper proportion, the proportion of the two planets' converging motions, and Jupiter's proper proportion. From proposition 8, 1:3 was composed of 1:2 plus the two proper proportions, so 1:3 minus 1:2 leaves 2:3 for the proper proportions.

This fifth is easily divided into its two components, the proper proportions of the motions of Saturn and Jupiter.

   11. *Proposition.* The proportion of the aphelial motion to the perihelial of Saturn had to be 4:5, a major third. That of the motions of Jupiter had to be 5:6, a minor third.

The reason for this is simple: there is no harmonic division of a fifth other than into major and minor thirds. By axiom 10 above, the third that is larger, and of the more masculine and excellent durus type of harmony, must have been given to the larger and higher planet Saturn; so the minor third is left for Jupiter.

---

[62] Here 1:2 and 1:3 "are the heads of the primary families of figures . . . by what was said in book 1." The first family includes figures (or quasi-figures) with 2, 4, 8, . . . , sides; the second family, figures with 3, 6, 12, . . . , sides. In book 1, see prop. 30, *G.W.* 6.34.

[63] At no. 13 in chap. 3, Kepler has noted in passing that the proportion of converging motions is greater than the proportion of the corresponding intervals. Ibid., 6.305.6–7.

In similar fashion, Kepler summarily extorts other harmonies from the polyhedral hypothesis:

> 12. *Proposition.* To Venus and Mercury was due the greater [diverging] harmony of 1:4, a double octave.

As Saturn and Jupiter are the outermost pair, so the innermost pair are Venus and Mercury, whose octahedron is from the same "marriage" as the outer pair's cube. The next proportion after 1:2 and 1:3 is 1:4. This is of the same family as 1:2 and commensurable with it, as the octahedron is of the same family and commensurable with the cube. Hence 1:4 belongs to Venus and Mercury.

Another, more specific reason for assigning 1:4 to the octahedral pair of planets is the square "hiding" within the octahedron, around its midsection. The square's proportion of spheres (a novelty, since the square is itself a plane figure) is $1:\sqrt{2}$, so the harmony of the octahedral pair is rightly obtained by doubling this proportion one or more times—twice, as it turned out.

The harmony 1:4 is that of the greater, that is, the diverging, proportion of the motions of this innermost pair of planets. This is archetypally well balanced, Kepler thinks, since 1:2 is the lesser proportion of the outermost pair. It is mathematically essential that 1:4 be the proportion of diverging motions, since it has been shown in chapter 3 that the proportion of converging motions must be smaller than the sesquialterate of the octahedron's proportion of spheres. A proportion of 1:3 would be double that proportion of spheres, and hence too large to be the proportion of converging motions, so 1:4 is far too large to be the proportion of converging motions. It must therefore describe the diverging motions.

Moving to Jupiter and Mars, Kepler produces an argument less convincing, perhaps, than any other in book 5.

> 13. *Proposition.* To the extreme motions of Jupiter and Mars were due these harmonies: a greater harmony of about 1:8, three octaves, and a lesser of 5:24, two octaves plus a minor third.

The ratio of spheres of the tetrahedron is twice that of the cube, so the proportions of Jupiter and Mars should ideally be 1:4 and 1:9, twice those of Saturn and Jupiter. But 1:4 is already taken, and 1:9 is not harmonic, so it has been necessary to use other proportions close to these. The closest neighboring harmonic proportions are 3:10 and 1:10, which arise from the pentagon, but that figure has nothing in common with the tetrahedron. Kepler settles on the next-closest harmonies for Jupiter and Mars, which are 5:24 and 1:8.

An easy calculation gives:

14. *Proposition.* The proper proportion of the extreme motions of Mars had to be made greater than a fourth, 3:4, and about equal to 18:25.

Subtracting the converging proportion of Jupiter and Mars, 5:24, and the proper proportion of Jupiter, 5:6, from the diverging proportion 1:8 leaves 18:25 as the proper proportion of Mars.

Next, Kepler deduces the remaining proportions of converging motions. These are particularly well defined, because in his initial hypothesis it is the converging distances that are governed by the regular polyhedra. A polyhedron occupies the space between the inner sphere of the outer planet (that planet's least distance) and the outer sphere of the inner planet (its greatest distance).

15. *Proposition.* The harmonies 2:3 (a fifth), 5:8 (a minor sixth), and 3:5 (a major sixth) had to be distributed, in the given order, among the converging motions of Mars and Earth, of Earth and Venus, and of Venus and Mercury.

Kepler has sound reasons, based on the regular polyhedra and the astronomical theory of chapter 3, for assigning these three harmonies to the converging motions. There are two basic arguments.

First, where two polyhedra have equal proportions of inscribed to circumscribed spheres, the corresponding pairs of planets should have almost equal proportions of converging distances. The cube's proportion of spheres equals that of the octahedron, so the proportion of Saturn's and Jupiter's converging distances should be close to that of Venus and Mercury. The dodecahedron's proportion of spheres equals that of the icosahedron, so the proportion of Mars's and Earth's converging distances should be close to that of Earth and Venus. This is a simple, sound geometrical argument.

Second, the proportions of converging motions depend not only upon the converging distances but also upon the eccentricities of the planets. Large eccentricities increase the disparity between the planets' proportion of distances and their proportion of converging motions. Kepler developed the theory behind this relation at the end of chapter 3, principally in his demonstration of the thirteenth proposition there. He cites it here in support of an informal lemma that he uses twice in the demonstration of this proposition 15:

195

[Lemma.] If two pairs of planets have the same proportion of converging distances, the pair with the greater sum of eccentricities necessarily has a smaller proportion of converging
motions.[64]

Kepler asserts the lemma in passing, merely claiming that it is true
"by Chapter 3." We can establish it easily from our discussion of proposition 13 in that chapter. For both pairs of planets, we had:

$$M_p:m_a = r_a:R_p + d \tag{9}$$

We know from the polyhedral hypothesis that $r_a:R_p$ is the same for
both pairs of planets, because the cube and the octahedron have the
same ratio of inscribed to circumscribed spheres. The current problem, then, is to show that the term d decreases if the sum of the
eccentricities increases. Equation (6) defined Kepler's "certain difference" d:

$$R_p:R + r:r_a = \tfrac{1}{2}r:R - d \tag{6}$$

Hence $d = \tfrac{1}{2}r:R - R_p:R - r:r_a$. Expressing $r:R$ as the sum of its
components,

$$d = \tfrac{1}{2}(r:r_a + r_a:R_p + R_p:R) - R_p: R - r:r_a$$
$$= \tfrac{1}{2}(-r:r_a + r_a:R_p - R_p:R)$$

But $r:r_a$ is the relative eccentricity of the inner planet, and $R_p:R$ is
that of the outer planet. If the sum of these eccentricities increases
while $r_a:R_p$ does not change, then d decreases; hence by (9) $M_p:m_a$
decreases, as was to be shown.

Kepler applies the lemma first to the converging motions of Venus
and Mercury. The proportion of their converging motions should be
close to the octave, 1:2, since that is the converging proportion of
Saturn and Jupiter, the planets whose interposed cube has the same
proportion of spheres as the octahedron between Venus and Mercury.
Suppose that it is in fact equal to 1:2. Subtracting 1:2 from Venus's and
Mercury's diverging harmony of 1:4 leaves 1:2 as the sum of their
proper proportions. But this does not equal the sum of the proper
proportions of Saturn and Jupiter, which is 2:3 by proposition 9.

It is easy to infer, qualitatively, what this discrepancy implies for the
converging proportion of Saturn and Jupiter. A planet's eccentricity, as
a proportion of its orbital radius, is very nearly half its proper proportion. Since the combined eccentricities of Venus and Mercury are
greater than those of Saturn and Jupiter, their proportion of converging motions has to be less than that for Saturn and Jupiter. Saturn's and

---

[64] Ibid., 6.337.8–9, 22–24.

Jupiter's proportion of converging motions has already been established as 1:2, so that of Venus and Mercury is close to this, but smaller. The next-smaller harmony than the octave is the major sixth, 3:5, so Venus and Mercury receive the major sixth for their proportion of converging motions.

Mars, Earth, and Venus, the planets whose orbits embrace the dodecahedron and icosahedron, are left to choose among the smaller consonances: the minor sixth, the fifth, and possibly the fourth; for the major and minor thirds have been ruled out by proposition 6. Kepler is able to use his lemma again, arguing that the dodecahedron received a smaller converging proportion than the icosahedron because Mars and Earth have a larger sum of eccentricities than Venus and Earth. The largest proportion among those remaining is the minor sixth, 5:8, so it belongs to the converging motions of Earth and Venus. To Mars and Earth are left the fifth, 2:3. (The next-smaller harmony, the fourth, is disallowed, by proposition 7, as a converging proportion unless the sum of the two planets' proper proportions is at least a fifth. Although Mars's proper proportion alone is nearly that large, Earth's is too small to make up the difference.)

Compared to its predecessor, proposition 16 is a simple calculation:

16. *Proposition.* The combined proper proportions of the motions of Venus and Mercury had to equal about 5:12.

The proportion of diverging motions is 1:4, by proposition 12, and the proportion of converging motions 3:5, by proposition 15, so the sum of the two proper proportions is the difference between these two, namely 5:12.

An evident consequence of proposition 16 is that the proper proportion of Mercury must be slightly smaller than 5:12. Kepler notes, however, that this holds true only by the logic of "these first reasons." Soon, he advises the reader, subsequent reasons would bring a little "yeast" to the common harmonies of Venus and Mercury, allowing the proper proportion of the latter to rise to exactly 5:12.[65]

The *rationes priores* close with a cautious attempt to fix the diverging harmony of Mars and Earth:

17. *Proposition.* A harmony of the diverging motions of Mars and Earth could not be less than 5:12.

[65] Ibid., 6.338.12–20.

197

The proportion of converging motions is 2:3, and the proper proportion of Mars 18:25. These two make a total of 12:25 as a lower bound for the diverging proportion, with Earth's proper proportion yet uncounted. The smallest harmonic proportion greater than 12:25 is 5:12. Hence "if a harmony is needed for this greater proportion of the motions of the two planets"—note the qualification—it would have to be a proportion no smaller than 5:12.[66] A great deal of further discussion will be required before Kepler is ready to explain (in proposition 42) why it was not possible for these two motions to receive a harmonic proportion.

Posterior Reasons: Consequences of the Universal Harmonies

In order to account for the last and most difficult of the heavenly proportions, those involving "obstructive Venus" and her metaphorical husband, Earth, Kepler makes a new start at this point in the chapter. The development of the *posteriores rationes*, as he called them, involves an immense effort. Kepler has already explained nearly all the proportions in the heavens by means of the *rationes priores*, using arguments that, if not totally conclusive, would at any rate leave most readers surprised at how thoroughly harmonic proportions permeated the heavens. Instead he insists upon explaining *all* the important proportions and upon addressing questions larger than the proportions implied by pairs of motions. And there is more to be gained in this attempt than a couple of recalcitrant proportions, for it is only in the study of Venus and Earth that the full richness of the celestial design is revealed. Investigating the dissonances, Kepler discovers a microtonal harmony more complex and more esoteric than he had suspected.

AXIOMS OF UNIVERSAL HARMONY

The posterior part of the chapter opens with four axioms noticeably more specific than the ones that came earlier:

18. *Axiom.* Universal harmonies of the motions had to be established, by the mutual accommodation of the six motions, and using especially the extreme motions.
This is shown by axiom 1.

---

[66] Ibid., 6.338.30–31.

19. *Axiom.* The same universal harmonies had to occur across a certain range of the motions, in order that they might thereby happen more often.

For if they had been limited to individual points of the motions it could have happened that they would never actually occur, or at least very rarely.

20. *Axiom.* Since the distinction of the types [genera] of harmony into durus and mollis is very natural, as was shown in book 3, universal harmonies of both types had to be obtained among extreme motions of the planets.

21. *Axiom.* Different instances of the harmonies of both types had to be established, so that the beauty of the world might be embellished by all the possible classes of variety; and this had to be done using at least some of the extreme motions.

By axiom 1.

The presence of universal harmonies, and several of each type, is a particular consequence of the general principle of variety. These harmonies, in turn, imply the existence of the scales that Kepler has already detected in the planetary motions:

22. *Proposition.* The extreme motions of the planets had to indicate the positions, or rather the strings, of the system of the octave, or the notes of a musical scale.

This is because the harmonic intervals based on a common note give rise to a scale, as had been shown earlier in book 3.

THE PAIR CHANGING THE TYPE OF HARMONY

Kepler continues with one of his most important propositions, one that is key to all of the *posteriores rationes* and that explains the very problematic harmonies that he has found between Venus and Earth.

23. *Proposition.* It was necessary that there be one pair of planets between whose motions there could be no harmonies other than the two sixths, the major 3:5 and the minor 5:8.

It is after all essential, by axiom 20, to distinguish the two types of harmony. Kepler thinks this task so important that the pair of planets devoted to it must not be distracted by any other harmonies: the two should devote their entire relationship to the distinction between durus and mollis. Since "the types of harmonies are distinguished by a diesis," which is the difference between the durus and mollis types of both thirds and sixths, little compass remains for variety in the proportions. The larger and smaller intervals must be nearly the same. In theory, the types of harmony can be distinguished by switching either between major and minor sixths or between major and minor thirds; and either alternative can be combined with one or more octaves. Kepler proves by exhaustion of possibilities that the sixths were necessarily used, and without supplemental octaves.

Thirds alone are impossible because of the polyhedra, as already demonstrated, since no regular polyhedron allows a ratio of spheres as small as a third. Furthermore, no third or sixth combined with any number of octaves has been found in the intervals determined thus far—except the diverging interval 5:12 of Mars and Earth, an octave plus a minor third. But that pair of planets is unsuitable. Its converging interval is the perfect fifth 2:3, a harmony that does nothing to change the type of harmony, and which is so much smaller than 5:12 that several intermediate and unwanted harmonies would occur in the transition.[67]

Only the sixths remain as possible termini for the proportions of a pair of planets devoted to changing the type of harmony. The durus and mollis sixths correspond to proportions of 3:5 and 5:8. For the difference in genus to be sufficiently marked, then, there must be a pair of planets whose extreme motions form these two proportions, which differ by a mere diesis. To sound these two intervals, but no others, both planets must have small eccentricities.

Obviously Kepler intends these leading roles for Earth and Venus, which gave him so much trouble in his earlier attempts to build universal harmonies. The roles here are so important that he analyzes them abstractly, and in considerable detail, through propositions 23–26, referring only to "the planets changing the type of harmony." Only in proposition 27 does he point out that the minor sixth has been given to Earth and Venus by the *rationes priores,* so that these must be the two planets in question.

> 24. *Proposition.* The two planets that change the type of harmony must make the difference of the proper proportions between their extreme motions a diesis; and they must make one sixth with the aphelial motions, the other with the perihelial.

[67] Ibid., 6.340.7–11.

This proposition comes as a surprise to the attentive reader. Up to this point the two harmonies of a pair of planets have arisen exclusively from their diverging and converging motions, so that the difference between those two harmonies has been the *sum* of the individual planets' proper proportions. Here in proposition 24 the diesis, which must be the difference between the two harmonies if the pair is to change the type of harmony, is assigned to the *difference* of the proper proportions. This can only be true, as noted, if the harmonies in question are not those between diverging and converging motions but rather those between aphelial and perihelial motions. For this one pair of planets only, the perceptive soul in the Sun attends the proportion of aphelial motions and the proportion of perihelial motions.

Kepler establishes this surprising result by ruling out any other way of obtaining the two sixths between the extreme motions of a pair of planets. If the proportion of diverging motions is a major sixth and that of converging motions a minor sixth—if, in other words, the usual two proportions are used—then either one planet's motion would vary by the entire diesis and the other not at all or each would vary by an interval less than a diesis. The first possibility, denying variety to one of the planets, would be in clear violation of axiom 4. The second possibility, almost as distasteful, would require that both proper proportions be smaller than the diesis, which is the least of the true *concinna*.

A third logical possibility, to which Kepler has been driven, is that for this one pair of planets, the two proportions of interest are not those of the diverging and converging motions but rather those obtaining when both planets are at aphelion, or both at perihelion. Such harmonies, which Kepler calls proportions of the motions *eiusdem plagae*, meaning "of the same region," have a sensuous advantage: they last longer. Before and after aphelion both planets are moving very slightly faster than their slowest motion, so that the proportion between them stays quite close to the major sixth for some time. Before and after perihelion both planets move very slightly slower than their fastest motion, so that again the proportion between them stays close to the minor sixth. Diverging and converging proportions do not remain harmonious for nearly as long, since the aphelial planet moves faster while the perihelial planet moves slower.[68] For the very important intervals that determine the genus of the chords of universal harmony,

[68] Ibid., 6.340.25–341.11.

Kepler thinks an extended duration important—apparently so that the harmony can better be enjoyed.

Kepler deduces the motions appropriate to the *planetae, genus Harmoniae mutantes* with great precision, by a succession of exquisitely detailed arguments.

> 25. *Proposition.* The superior of the planets changing the type of harmony had to have a proportion of proper motions less than a minor tone, 9:10; the inferior, less than a semitone, 15:16.

The proper proportion of the inferior of the pair had to be less than a semitone, 15:16. Otherwise the additional interval of a perfect fifth would occur between the diverging motions of this pair. This is a simple calculation: if the perihelial motions make a minor sixth, 5:8, and the proper proportion of the inferior planet makes a semitone, 15:16, then the converging motions, when the outer planet was at perihelion and the inner planet at aphelion, would make a proportion of $\frac{5}{8} - \frac{15}{16} = \frac{2}{3}$, a perfect fifth. The occasional occurrence of this harmony might distract the two planets from their assigned roles as planets changing the type of harmony.

From this it follows that the proper proportion of the superior of the pair, which is known to be a diesis greater than the proper proportion of the inferior of the two, must be less than a semitone plus a diesis, or a "minor tone." Numerically, $\frac{15}{16} + \frac{24}{25} = \frac{9}{10}$.

Reconstructing what even he admits are very delicate choices that the Creator must have made among tiny, unpleasant intervals, Kepler finally decides that the magnitudes of the two planets' proper proportions can be justified *exactly*:

> 25. *Proposition.* The superior of the planets changing the type of harmony had to have for the interval of its extreme motions either two dieses, 576:625, or about 12:13; or a semitone, 15:16; or something intermediate, differing from one or the other by a comma, 80:81. The inferior had to have either a simple diesis, 24:25; or the difference between a semitone and a diesis, 125:128 or about 42:43; or else similarly something intermediate, differing from one or the other by a comma, 80:81. And indeed the former [must be] a double diesis, the latter a simple diesis, each interval diminished by a comma.

Some discussion is required to establish this result. The proper interval of the superior planet has to be greater than a diesis, by proposition 24, but less than a minor tone, by proposition 25.[69] Both proper intervals should, if possible, be pleasing (*concinna*), although they are too small to be actual consonances. But the only pleasing intervals smaller than the minor tone are the semitone and the diesis, and the difference between these is only 125:128, smaller than the required diesis. It is not possible, therefore, that both proper proportions are *concinna* intervals.

The next-most harmonious possibilities, Kepler continues, are either that the superior planet has a proper interval of a semitone, and the inferior planet a diesis less than a semitone, which is $\frac{125}{128}$; or that the proper intervals are respectively two dieses and one diesis. If it is true that unpleasing intervals cannot be avoided, then the two proper proportions should be equally distant from a pleasing interval, "since [the proper proportions] of each planet have equal rights."[70] Their distance from a *concinna* interval should be a comma, 80:81, a tiny interval that appears, in the heavens as in music, only as the difference between larger intervals.

As to which of the possible pairs of pleasing intervals the proper proportions should be based on (semitone and semitone-minus-diesis, or double diesis and diesis), Kepler judges that "the dieses win the argument." The semitone is a regular part of the scale, but its companion 125:128 is not. The diesis likewise is not in the scale as a distinct interval. It figures prominently as a component in the analysis of the system of the octave in table 7.2, however, and the double diesis figures also, in a way, since the decomposition of the octave does yield two dieses. The diesis, moreover, was the original criterion in this business of distinguishing durus from mollis, and hence fits comfortably into the harmonic niches of the theory.[71]

Kepler concludes that the proper proportion of the superior planet

[69] Kepler cites prop. 25 for the claim that the superior planet needed a proper proportion greater than a diesis (p. 225.11–12 in the 1619 edition; *O.O.* 5.308.7–8; *G.W.* 6.342.3–4). But this follows from prop. 24, not prop. 25. Since prop. 24 is cited correctly in the next sentence, this was not a simple error on Kepler's part. It is possible that a twenty-fifth proposition (or corollary) originally stated something along the lines that "the superior of the planets changing the type of harmony has a proper proportion greater than a diesis," and that Kepler later deleted this proposition as trivial.

[70] *Cum vero utriusque Planetae sint aequalia jura*, *G.W.* 6.342.12. Coming from Kepler this argument sounds forced, since the Earth was not only the superior planet but the metaphorical husband.

[71] Ibid., 6.342.25–35.

is indeed best placed at exactly two dieses minus a comma, or 2,916:3,125 (about 14:15); and that of the inferior planet at a single diesis minus a comma, or 243:250 (about 35:36). The difference between these two intervals is exactly a diesis, so the planets would indeed be able to change the type of the universal harmony by moving between their aphelial and perihelial intervals. Needless to say, these quasi-pleasing harmonies agree closely with the proper proportions arising from the known eccentricities of Earth and Venus.

It might be asked, Kepler sagely continues, whether the highest creative wisdom really occupied itself with such trivial logic-chopping (*tenuibus istis ratiunculis*). He himself is content to believe this: first, because many other reasons might exist that have escaped him; and second, because God has established nothing without good reason. It would be much more absurd, he is certain, to suppose that these tiny quantities had been assigned without a rational cause: "It is not enough to say that God used proportions of a size that pleased him: for in geometrical matters subject to free choice, nothing has pleased God without some kind of geometrical cause—as can be seen in the borders of leaves, the scales of fish, the hides of wild animals, and their spots and arrangement of spots, and such things."[72]

Given such detailed knowledge of the astronomical attributes of the *planetae, genus Harmoniae mutantes*, it is easy to identify which they are.

27. *Proposition.* The larger proportion of the motions of Earth and Venus had to be the major sixth, between the aphelial motions; the smaller, the minor sixth, between the perihelial motions.

Kepler could have termed this a corollary, but for its importance. To justify it he cites axiom 20 and propositions 15, 23, and 24. This result explains, he remarks, why the exact harmonies for Earth and Venus are found between the aphelial and perihelial motions rather than between the converging motions as was true for the superior planets.

28. *Proposition.* To Earth a proper proportion of motions of about 14:15 applies; to Venus, about 35:36.

These have already been calculated in proposition 26. In chapter 4 Earth's observed proportion was given as 14:15, and Venus's as 34:35, which is hardly distinguishable from 35:36.

[72] Ibid., 6.342.36–343.6.

Next Kepler turns back to the other planets, to establish the reasons behind the rest of the harmonic structure underlying the great chords he exhibited in chapter 7.

29. *Proposition.* The greater harmony of the motions of Mars and Earth, that of the diverging motions, could not be among those harmonies greater than 5:12.

This one is subtle. Proposition 17 has already established that this proportion is not a harmony less than 5:12. (To recall the argument, the proportion of converging motions and the proper interval of Mars sum to something more than 12:25. The diverging proportion, which includes these and Earth's proper proportion as well, must be at least this large; and the smallest harmony greater than 12:25 is 5:12.) To show that the diverging proportion is not a harmony greater than 5:12, Kepler now adds in the last component, Earth's proper proportion of about 14:15, and obtains "more than 56:125" as a nearly exact result. The next harmony greater than 56:125 is still 5:12, an octave plus a minor third, so the proportion of Jupiter's and Mars's diverging motions could not be a harmony less than that. "Note," Kepler adds after this calculation, "that I do not say that this proportion is neither larger nor smaller than 5:12. But I do say this: if it is necessary that it be harmonic, no other harmony is suitable for it."[73]

The inequality ("more than 56:125") admitted in this calculation, and suggested in proposition 17, is typical of the thicket of cross-references making up chapter 9. Formally, the inequality is concessive, since the present proposition would hold all the more if the calculation yielded exactly 56:125. Yet Kepler has reason for granting the excess. The proportion of Jupiter's and Mars's diverging motions is greater than 56:125 because the proper proportion of the motions of Mars was greater than 18:25 in proposition 14. The calculation there yielded 18:25 exactly, with but a hint that the true value would increase if the diverging proportion of Jupiter and Mars were to grow beyond 1:8. It will grow, but only in proposition 40, and only by a Platonic limma, 243:256. Adding in the Platonic limma here by anticipation would yield 1,701:4,000, to which 5:12 is still the next-greater harmony.

30. *Proposition.* The proper proportion of the motions of Mercury had to be made greater than all the other proper proportions.

[73] Ibid., 6.344.5–7.

This is almost obvious. By proposition 16 the sum of the proper proportions of Venus and Mercury has to be 5:12, and by proposition 26 that of Venus is 243:250. This leaves Mercury alone a proper proportion of 625:1,458, considerably larger than that of any other planet.

Kepler pauses here to observe that the proper proportions of the four outer planets sum to 56:135, which very nearly equals 5:12, the sum of the proper proportions of Venus and Mercury. "This indeed was not sought," he points out, "nor was it chosen from some distinct and particular archetype of beauty; but it happened as a bonus, by necessity of the related causes from the harmonies thus far established."[74]

Kepler now cues the entry of the bass into the harmonic framework he has so laboriously defined for the two genera.

31. *Proposition.* The aphelial motion of Earth needed to agree with the aphelial of Saturn across several octaves.

For the harmonies to be universal, Saturn must somehow harmonize with Earth and Venus. These harmonies could be either on the same note across octaves (*identisonus*), or on different notes (*diversisonus*) that are in harmonic proportion. Both could not be harmonies on different notes, however: the motions of Earth and Venus, when in harmony, form a major or minor sixth, and neither sixth can be harmonically divided by two notes (the aphelial and perihelial motions of Saturn) into three intervals. One of Saturn's motions therefore has to sound the same note, lowered of course by several octaves, as an extreme motion of either Earth or Venus.

Agreement of identical notes is more excellent, Kepler continues, than that of different notes. Hence the identical agreement across octaves has been assigned to the more excellent of the harmonies of Earth's and Venus's extreme motions. The aphelial motions of Earth and Venus are more excellent for two reasons: because of their greater elevation from the Sun (alluding to axiom 10), and because their aphelial harmony is durus, which is the more excellent type of harmony. The aphelial motion of Saturn, then, should agree identically across the octaves with the aphelial motion of Earth, the higher of the pair of planets distinguishing the type of harmony.

This conclusion also agrees better with the less intricate calculations of the *rationes priores.* "The posterior reasons, with which we are

---

[74] Ibid., 6.344.22–24.

dealing now, modify the prior reasons, but only regarding small amounts, which in Harmony means regarding intervals smaller than all the *concinna*." In fine-tuning the harmonic structure that he has built upon the prior reasons, Kepler does not want to destroy it, and those early propositions already imply that Earth's aphelial motion is more likely than that of Venus to sound the same note as Saturn's aphelial motion. For the proportion between the aphelial motions of Saturn and Earth can be decomposed into the proper proportion of Saturn (4:5, by proposition 11), plus the converging proportion of Saturn and Jupiter (1:2, by proposition 8), plus the diverging proportion of Jupiter and Mars (1:8, by proposition 13), plus the converging proportion of Mars and Earth (2:3, by proposition 15), making a total of 1:30. Augmenting this by a mere semitone, 15:16, makes it exactly five octaves, to the great enhancement of the universal harmonies. The adjustment can be divided into tiny pieces, each much less than the smallest of the *concinna*, and distributed as seems prudent among the intervening intervals established by the prior reasons.

32. *Proposition.* In the planets' universal harmonies of the mollis type, the precise aphelial motion of Saturn could not agree exactly with the other planets.

Since Earth's aphelial motion, and the major sixth it forms with Venus, creates the durus harmony, neither it nor the identically sounding note of Saturn's aphelion can enter exactly into the mollis harmony. There is a slightly faster motion of Saturn, close to its aphelion, Kepler notes, that does enter into the mollis harmony—as has indeed been seen in chapter 7.[75]

The next two propositions are again abstract, continuing the series of propositions 22–26. They are even in the present tense.

33. *Proposition.* The durus type of harmonies and of the musical scale is closely related (*familiaris*) to the aphelial motions; the mollis, to the perihelial.

Admittedly it is not the aphelial motion alone of Earth but rather a range of its motions, that can make a major sixth with some motion of Venus; and not its perihelial motion alone that can make a minor sixth.

[75] Ibid., 6.346.1–6. In the five-planet skeletons of chap. 7, mollis as well as durus (table 9.10), Saturn's G in the bass arises from a motion of 1′47″, quite close to its aphelial 1′46″.

Yet the permanent and clear designation of the type is made only by extreme motions.

34. *Proposition.* The durus type is closely related to the superior planet in a comparison of two; the mollis type to the inferior.

Just as the durus type of harmony is related to the aphelial, slowest motion of a single planet, so it is related to the slower-moving of a pair of planets.

Kepler is now able to characterize the two outer planets with respect to the two types of celestial harmony:

35. *Proposition.* Saturn and Earth embrace the durus type more closely; Jupiter and Venus, the mollis type.

Saturn's aphelial motion sounds the same note, several octaves lower, as Earth's aphelial motion; so the former is involved in the durus harmony designated by the latter. Saturn "cherishes" the durus type, by proposition 31, and "spurns" the mollis type, by proposition 32. Jupiter is inferior to Saturn, so to it is due the mollis type of harmony, by proposition 34.

Together, propositions 33 and 35 imply that the characteristic tuning for durus harmonies is one containing the aphelial motions of Saturn and of Earth, a tuning that yielded the lowest-tuned chord skeletons on both sides of table 9.8 and both sides of table 9.10. The characteristic tuning for mollis harmonies is one containing the perihelial motions of Jupiter and Venus, a tuning that yielded the highest-tuned skeletons on both sides of table 9.10. As chord skeletons, indeed, the latter do not include Venus's perihelial motion, because E is not in the chord with G, B or B♭, and D. Yet all the notes, including Venus's E, are tuned to a system that includes the two motions most strongly associated with mollis harmony.

36. *Proposition.* The perihelial motion of Jupiter had to agree with the perihelial motion of Venus in the same musical scale, but not in the same harmonic interval; much less could it be in a harmonic interval with the perihelial motion of Earth.

The perihelial motion of Jupiter, as the motion associated with mollis harmony, needs to mark a definite note or tone of the mollis scale. The perihelial motions of Venus and Earth make that scale possible; hence Jupiter's perihelial motion must join in the same tuning with them.

It cannot be in harmonic proportion with the perihelial motion of Venus, however. Venus's *aphelial* motion is a major sixth above that of Earth, and hence a major sixth and some octaves above the aphelial motion of Saturn. If Saturn's aphelial motion is G, Venus's aphelial motion is thus E some octaves above. But Jupiter's perihelial motion is an octave and a fifth above Saturn's aphelial motion, by proposition 8, so its note is D. Venus's aphelial E is only a minor tone 9:10 above Jupiter's perihelial D, which is much less than the smallest harmonic interval 5:6; and Venus's perihelial motion is less than a diesis higher than the aphelial E. Thus it is mathematically impossible for Jupiter's perihelial motion to be in harmony with both the aphelial motion of Saturn and any motion of Venus.

In figure 9.5, which is a mollis scale that has been tuned with Saturn's perihelial motion at G, the perihelial motions of Jupiter and Venus play B♭ and C. As noted thereafter under "The Genera of the Scales," the aphelial motion of Venus (though absent from the figure) is actually closer to the note C than the perihelial motion is. Kepler forced the perihelial motion into figure 8.4, evidently for the sake of this contrast between aphelial motions of Saturn and Earth as durus, and perihelial motions of Jupiter and Venus as mollis.

Kepler concludes his discussion of proposition 36 by showing why Jupiter's perihelial motion is unable to form a harmonic proportion with the perihelial motion of Earth, which also plays an important role in mollis harmony. Earth's perihelial motion is a minor sixth (5:8) lower than Venus's perihelial motion, which is slightly higher than E; and hence is very nearly a minor sixth lower than Venus's aphelial E. Jupiter's D is only a minor tone (9:10) lower than E, so it is higher than Venus's perihelial motion by a bit more than 5:8 − 9:10, or 25:36. Such an interval, equal to two minor thirds, or a fifth minus a diesis, is decidedly unharmonic.

BLEMISHES

This was as far as Kepler could go in fitting the planetary motions to harmonies, and by implication as far as the Creator had been able to go in establishing the motions in harmonic proportions. The competing claims of different proportions to be harmonic were, Kepler realized, impossible to accommodate perfectly. He had adjusted the diverging proportions of all pairs except Earth and Venus (whose diverging proportion was not used), and the proper proportions of several planets, in fitting them together as harmoniously as possible. Most of the rest of chapter 9 acknowledges these remaining defects in the structure, and explains how they were unavoidable, and indeed were the least offensive that could possibly have emerged.

Propositions 37–43 explain some small but excruciatingly unpleasant adjustments to the harmonies Kepler established earlier in the *rationes priores*.

> 37. *Proposition.* An interval equal to the [proper] interval of Venus had to be added to the sum of the proper harmonies of Saturn and Jupiter, 2:3, and to their larger joint harmony, 1:3.

This adjustment (Kepler calls it the *additamentum*) is needed to resolve an apparent inconsistency that emerged from the discussion of the previous proposition. Let us use V_a and V_p for the extreme motions of Venus, and δ for the tiny proportion between them. Then δ is smaller than a diesis, and $V_p = V_a + \delta$. We cannot simultaneously have:

> A) Jupiter's perihelial motion J_p is exactly an octave and a fifth (1:3) higher than Saturn's aphelial motion S_a, as in proposition 8. With S_a at G, J_p is at D.
> B) Venus's aphelial motion V_a is a major sixth above Earth's aphelial motion (proposition 27), hence a major sixth above S_a (proposition 31). With S_a at G, V_a is at E.
> C) Venus's perihelial motion V_p (or $V_a + \delta$) is in the same scale with J_p (proposition 36). With S_a at G, V_p is less than a diesis above E but lower than F. It cannot be in the same scale with J_p at D.

The interval between Venus's two extreme motions must be added, therefore, to the proportion of Jupiter's and Saturn's diverging motions, raising the pitch of J_p. Putting J_p at the pitch D + δ allows it to be in the same scale with V_p at the pitch E + δ. This mollis scale would of course not be the same scale as the durus scale wherein S_a and E_a are at G, and V_a at E.

The proportion of Jupiter's and Saturn's diverging motions has three components: the two proper proportions and the proportion of converging motions. The last of these proportions is precisely 1:2 by proposition 8, so the increment must accrue to the sum of the proper proportions.

The theoretical diverging proportion now amounts to 1:3 plus 243:250; yet the proportion calculated from "observed" motions in chapter 4 was still larger, 1:3 plus about 26:27 or 27:28. An additional second of arc in Saturn's daily motion at aphelion—which, Kepler asserts, could scarcely be discerned—would bring the observed proportion into line with the regrettably unharmonious result of this adjustment.

Kepler continues, relentlessly, dividing the *additamentum* among the proper proportions of Saturn and Jupiter.

> 38. *Proposition.* The *additamentum* 243:250 to the sum of the proper proportions of Saturn and Jupiter, which until now by first reasons was set at 2:3, had to be distributed thus between the planets: to Saturn accrued a comma, 80:81, to Jupiter the remainder 19,683:20,000 or about 62:63.

It is necessary to divide the *additamentum* between the proper proportions of the two planets, Kepler argues, rather than give all of it to either of them, so that each could agree "with some latitude" in the universal harmonies of its own type, durus for Saturn, mollis for Jupiter. He cites proposition 19, which stated that the universal harmonies must hold within some range (*latitudo*) of motions, so that they can occur more often. It is not obvious, however, why small increases in both proper proportions provide more of a range for universal harmony than an increase in only one of the proper proportions.

At any rate, the amount to be divided was smaller than all the agreeable intervals. Only one way of dividing it suggests itself to Kepler—a way to which he earlier resorted when allocating proper intervals to Earth and Venus in proposition 26. A comma should go to one planet, and the remainder to the other. The remainder in this case is 243:250 − 80:81, or 19,683:20,000—very nearly 62:63, which is thus a larger portion than the comma. With a careful eye on axiom 10, Kepler nevertheless assigns the comma to Saturn, "the higher and more worthy planet" and the planet possessed of the larger proper proportion. For the comma is more harmonic than the remainder.

With this division of the *additamentum*, the proper proportion of Saturn becomes 64:81, an "adulterated major third."[76] That of Jupiter becomes 6,561:8,000, an irreducible proportion that one is tempted to call a defiled minor third. The revised proportion of Saturn's motions implies for it a proportion of distances equal to 8:9, since that propor-

---

[76] The adulterated intervals, differing by a comma from their pure counterparts, could be derived from the "ditone," two consecutive major tones that naturally occurred between F and A or between E and G♯. See book 3, chap. 12 (ibid., 6.156–57).

tion is precisely half of 64:81. Kepler admits that he is uncertain whether this small harmony in the distances is a cause or a result of the corrected harmony in the extreme motions.[77]

After adjusting the proper proportions of the outer two planets, Kepler cannot hope to bring all four of their extreme motions into any kind of universal harmony. He retains those essential to the type of harmony associated with each planet: the aphelial motion of Saturn and the perihelial motion of Jupiter. The other motions are necessarily out of tune:

> 39. *Proposition.* In the universal harmonies of the planets of the durus type, Saturn could not agree with its exactly perihelial motion, nor Jupiter with its exactly aphelial motion.

Saturn's aphelial motion agrees with the universal durus harmony; so its perihelial motion, an adulterated major third higher, cannot. And Jupiter's aphelial motion is exactly an octave higher than Saturn's perihelial motion, since the proportion of converging motions has been left pristine at an octave.

> 40. *Proposition.* It was necessary that a Platonic limma be added to the three-octave (1:8) harmony of the diverging motions of Jupiter and Mars, as established by the prior reasons.

In his discussion of proposition 31 Kepler calculated that prior reasons left the harmony of Earth's and Saturn's aphelial motions just a semitone short of five octaves, and went on to conclude that the harmony really was exactly five octaves. Of the semitone needed to accomplish this adjustment, a comma is now allocated between Saturn's aphelial and perihelial motions. This leaves $15:16 - 80:81 = 243:256$, a Platonic limma, to be inserted, perhaps piecemeal, somewhere between Saturn's perihelial motion and Earth's aphelial motion. The converging proportions of Saturn and Jupiter and of Earth and Mars are not to be touched; Kepler consistently treats the converging proportions of the superior planets from the prior reasons as exact. There remains only the interval between Jupiter's aphelial motion and Mars's perihelial motion to accept the Platonic limma.

The harmony of the diverging motions of Jupiter and Mars thus became $1:8 + 243:256 = 243:2{,}048$. Kepler notes that this proportion is intermediate between 1:9 and 1:8, "of which an earlier analogy re-

<hr/>

[77] Ibid., 6.348.37–349.3.

quired the former, while harmonic agreement is closer to the latter."[78] The allusion is to the discussion of proposition 13. Since the tetrahedron's proportion of spheres (1:3) is double that of the cube (1:$\sqrt{3}$), it seemed to Kepler there that the diverging proportion of the planets embracing the tetrahedron should be double that of the planets embracing the cube. Saturn and Jupiter's diverging proportion was 1:3 at that point, so this analogy suggested 1:9 for Jupiter and Mars.

41. *Proposition.* The proper proportion of the motions of Mars was necessarily made twice the harmonic proportion 5:6, that is, 25:36.

Remarkably enough, this is nothing but a calculation.
$$M_a{:}M_p = J_a{:}M_p - J_a{:}J_p - J_p{:}M_a$$
These three components have been determined in propositions 40, 38, and 13, respectively:
$$M_a{:}M_p = 243{:}2048 - 6561{:}8000 - 5{:}24$$
$$= 25{:}36$$
The cascade of adjustments compresses Mars's and Earth's diverging motions into a dissonance:

42. *Proposition.* The greater or diverging common proportion of Mars and Earth was necessarily made 54:125, less than the harmony 5:12 established by the prior reasons.

The components of this diverging proportion were determined by propositions 41, 15, and 26, respectively.
$$M_a{:}E_p = M_a{:}M_p + M_p{:}E_a + E_a{:}E_p$$
$$= 25{:}36 + 2{:}3 + (24{:}25 + 24{:}25 - 80{:}81)$$
$$= 54{:}125$$
This is less than the prior harmony 5:12 by a very small amount, less than any pleasing interval.[79]

The aphelial motion of Mars has given trouble all along. In chapter 7 Kepler was unable to fit it into a chord until he dropped both Venus and Earth from the chord skeletons. It is a tone or less away from being two octaves lower than any motion of Venus, and well within a minor third (the smallest consonance) from being one octave lower than any motion of Earth. Kepler finally spells out these difficulties in proposi-

---

[78] Ibid., 6.349.34–37.

[79] At ibid., 6.350.36–37 (twice), the proportion 608:1,500 should be 648:1,500, an error in the 1619 edition missed by Frisch and Caspar; and the approximation to this should be 5:12 minus 27:28, rather than 5:12 minus 36:37. (The erroneous 608:1,500 is *greater* than 5:12 by about 36:37.)

tion 43, showing that Mars's aphelial motion is necessarily unsuited for any universal harmony, although it can at least sing one of the tones in the mollis scale.

43. *Proposition.* The aphelial motion of Mars could not really agree with any universal harmony; yet in a way it had to come into harmony with a scale of the mollis kind.

In fact, the perihelial motion of Jupiter, 5′30″, was at the note D in some of the characteristic tunings (for example, the highest tunings in table 9.10). Proposition 13 placed the aphelial motion of Mars two octaves plus a minor third (5:24) higher than this. That proportion, like all the converging intervals from the *rationes priores*, has not been changed by subsequent propositions. A minor third above D, however, is not F in Kepler's system, but rather is a comma higher than F.

To explain this, Kepler inserts here a supplement to his discussion of "adulterated consonances" in chapter 12 of book 3, which had evidently been typeset before he wrote these late propositions. There he listed the adulterated intervals, just a comma too large or too small, that are found only when the system of the octave is doubled by placing one octave above another. He now points out that such imperfect consonances lurk even within a single octave. From D to F is a minor tone and a semitone, making an interval of $9:10 + 15:16 = 27:32$, precisely a comma less than the minor third 5:6. Since Mars's aphelial motion is a full, unadulterated minor third (plus two octaves) above Jupiter's perihelial D, it is a comma too high to be F. Mars's aphelial motion thus does not participate fully in the mollis scale, but at least it participates to a certain extent.

Even in this mollis tuning, however, it is not consonant. The perihelial motion of Venus is at E, which forms a dissonance equal to a semitone plus a comma with the adulterated F. The aphelial motion of Venus is lower by a diesis minus a comma, so it forms the interval of a minor tone (a semitone plus a diesis) with Mars's aphelial motion. Hence no motion of Venus makes a consonance with the slowest motion of Mars in the celestial mollis scale.

The durus scale characteristically (that is, in the tunings that include the aphelial motions of Saturn and Earth) places the aphelial rather than the perihelial motion of Venus at E. The aphelial motion of Mars is a minor tone higher than E. This leaves it between F and F♯, making a very unpleasant interval of 25:27 with G.

Kepler calculates also that none of Earth's motions can harmonize with the aphelial motion of Mars. The necessity of durus and mollis harmonies, and the consequent existence of Earth and Venus as planets changing the type of harmony, have implied a structure in which the aphelial motion of Mars cannot harmonize. It is no wonder that, in chapter 7, he was unable to work this last extreme planetary motion into his chord skeletons until he dropped both Earth and Venus from them.

44. *Corollary.* It is evident therefore—from this proposition 43, regarding Jupiter and Mars, and from proposition 39, regarding Saturn and Jupiter, and from proposition 36, regarding Jupiter and Earth, and from proposition 32, regarding Saturn—why above in chapter 5 the extreme motions of the planets could neither all be accommodated perfectly to any one natural system or musical scale; nor to ones which were accommodated to a system of the same tuning. They all were found to have divided the notes by the natural reason of that [system], or to have formed a purely natural succession of pleasing intervals. For prior to this system were the causes by which individual planets obtained individual harmonies; by which all the planets obtained universal harmonies; and by which there were two types of universal harmonies, durus and mollis. Given these things a complete accommodation to a single natural system was already blocked. But if those causes had not necessarily taken precedence, there is no doubt that a single system and a single tuning would have embraced the extreme motions of all the planets. Even if there had been need for two systems, corresponding to the two types of song, durus and mollis, the same natural order of the scale would have been expressed not only in the one system of the durus type but also in the other of the mollis type. I reiterate that you have here, therefore, the causes promised in chapter 5 of the small disagreements, all of them less than the *concinna* intervals.

This is not merely a summary. Kepler claims that the attainment of any single universal system of harmony was blocked because of the priorities among the different types of harmony. Harmonies of single planets, universal harmonies joining the different planets, and the realization of the two different types of harmony are more important overall than the joining of all the harmonies into a unified system. In the eternal mathematics of the Creator, it turned out that a harmony so fertile was incapable of complete unification.

In the *rationes priores*, the sum of the proper proportions of Mercury's and Venus's motions was placed at 5:12, with a promise that the proper proportion of Mercury would eventually be given "some yeast," which would raise it alone to 5:12. The yeast is now available.

45. To the larger shared interval of Venus and Mercury, a double octave, and also to the proper interval of Mercury, as established in propositions 12 and 16, there had to be added an interval equal to the interval of Venus. Thus the proper interval of Mercury became exactly 5:12, and both motions of Mercury agreed with the perihelial motion of Venus.

Kepler justifies this tiny and precise correction at length. He begins, characteristically, with a broad analogy:

For since the aphelial motion of Saturn had to agree with the aphelial motion of Earth—that of the outermost planet, circumscribed to its figure, and highest, with the highest motion of Earth, which distinguishes the classes of figures—it follows by the laws of opposition that the perihelial motion of Mercury would agree with the perihelial motion of Earth—that of the innermost planet, inscribed to its figure, and lowest, and closest to the sun, with the lowest motion of Earth, that common boundary: the former designating the *durus*, the latter the *mollis* type of harmonies.[80]

If the perihelial motion of Mercury harmonizes with the perihelial motion of Earth, as this analogy indicates, then it also has to harmonize with the perihelial motion of Venus, since the latter is exactly a minor sixth higher than that of Earth. The double octave from prior reasons (see table 9.3) therefore belongs not to the diverging, but to the perihelial motions of Venus and Mercury. Mercury's apsidal motion is faster than the *rationes priores* implied, by exactly the proper proportion of Venus's motions.

This solves the problem of Mercury's proper proportion. The converging harmony of Venus and Mercury is exactly 3:5 by proposition 15, so

$$Me_a{:}Me_p = V_p{:}Me_p - V_p{:}Me_a$$
$$= 1{:}4 - 3{:}5$$
$$= 5{:}12$$

---

[80] Ibid., 6.352.32–38. The classes of figures, male and female, were distinguished in chap. 1 (ibid., 6.292.9–17). The male figures, with the androgynous tetrahedron, are superior to Earth, while the female figures are inferior to Earth. Hence Earth's orbit is a *sepes*, a fence or boundary.

Proposition 15 itself was hardly a model of exact reasoning, however. Kepler simply argued by analogy that this harmony needed to be close to 1:2, but smaller, and took the next smaller consonance, 3:5.

Kepler provides a scattering of other arguments here. In the analogies of book 5, chapter 1, Mercury alone played a role corresponding to Saturn and Jupiter together. The two outermost planets were entirely outside—and did not at all touch—the figures of the dodecahedral marriage, namely the dodecahedron and the icosahedron. On the inside, only Mercury failed to touch those figures. Since the combined proper proportions of Saturn and Jupiter have been augmented by an amount equal to Venus's proper proportion in proposition 37, it is fitting that Mercury's proper proportion be increased by a like amount.[81]

A third argument urges that just as the slowest motion in the planetary system is exactly five octaves from the slowest motion of the *planetae, genus Harmoniae mutantes,* so the fastest motion in the system should be some exact number of octaves from the fastest motion of those two planets.[82]

Fourth, the three outermost planets have one motion each that participates in the universal harmony. Next come Earth and Venus, the planets that change the type of harmony. Since only one planet is inferior to them, both its motions should play a part in the universal harmony. Kepler condensed this argument into a single sentence, but I think it can be construed. Each of the three superior planets has one extreme motion that is in the scales and one that is not. Saturn's aphelial motion and Jupiter's perihelial motion are fundamental to the durus and mollis harmonies, respectively. (Their other extreme motions are not in the universal harmonies. Saturn's perihelial motion is an adulterated major third—a major third plus a comma—higher than its aphelial motion. Jupiter's aphelial motion is exactly an octave higher than Saturn's perihelial motion, and hence is excluded also.) As for Mars, its perihelial motion is in the harmony, exactly a fifth lower than Earth's aphelial motion, and hence playing C in the durus scale. Its aphelial motion is dissonant in the durus scale and plays only an adulterated note in the mollis scale, by proposition 43.

Since the three outer planets have one extreme motion each in the universal harmony, Kepler apparently believes that Mercury, as the

---

[81] Ibid., 6.353.8–17.
[82] Ibid., 6.353.18–24.

sole planet inside the pair changing the type of harmony, needs to have both its extreme motions in the universal harmony, for balance.[83]

Finally, Kepler points out that among the three pairs of adjacent planets exterior to Earth it has been the converging harmonies that are exact, while the diverging and proper harmonies have suffered small adjustments. Among the two pairs of adjacent planets interior to Earth, it is the proportions not of converging or diverging motions but rather of motions *eiusdem plagae*, aphelial with aphelial and perihelial with perihelial, that have received the exact harmonies. The most important pair, Earth and Venus, have the major and minor sixths. The other pair, Venus and Mercury, require two exact harmonies. But since they are not the pair changing the type of harmony, they do not require exact harmonies between both their aphelial and perihelial motions: hence they have received exact harmonies between perihelial motions and between converging motions.[84]

Kepler is grasping at the Creator's reasons here, heaping one imperfect analogy atop another. He is at least able to exhibit some evidence of design in the proportions inferior to Earth. Venus has the smallest proper proportion of all the planets, and the most unharmonic; Mercury has the largest, and the most precisely harmonic: an absolute harmony, without leavening.[85]

Pleased with the symmetries he has discovered among the harmonies, Kepler turns his attention back to the polyhedra for a while.

46. *Axiom.* The interposition of solid figures between the planetary spheres, if it is free, and not impeded by the necessities of preceding causes, must follow completely the analogy of geometrical inscriptions and circumscriptions; and thus also the conditions of the proportion of inscribed and circumscribed spheres.

47. *Proposition.* If the inscription of figures between the planets had been free, the tetrahedron would have had to touch exactly with its angles the perihelial sphere of Jupiter above, and touch exactly with the centers of its sides the aphelial sphere of Mars below. The cube and the octahedron, with their angles resting on the perihelial sphere of their respective planets, would have had to penetrate the spheres interior to them with the centers of their sides, so that those centers would be located between the aphelial and perihelial spheres. The dodecahedron

---

[83] Ibid., 6.353.24–27.
[84] Ibid., 6.353.27–38.
[85] Ibid., 6.354.2.

and icosahedron, on the other hand, while just touching with their angles the perihelial spheres of the planets outside them, would not quite have reached the aphelial spheres of the planets interior to them with the centers of their sides. Last, the dodecahedral hedgehog, with its angles resting in the perihelial sphere of Mars, had to approach closely the aphelial sphere of Venus with the midpoints of the turned-in edges, connecting two solid spikes.

The turned-in edges (*latera inversa*) of a hedgehog are just what is described here: the edges connecting two of the solid angles or "spikes." On figure 9.2 he marked them with the letter *O*.[86] If one looks closely, one will see that they are the edges of the regular dodecahedron from which the hedgehog's spikes have sprouted.

In the viewpoint Kepler adopts for proposition 47, a polyhedral figure stands on its solid angles, as if on so many pointed feet, resting upon the inner sphere of the planet outside it. With its faces it embraces, more or less closely, the outer sphere of the next planet inward. The nesting is not perfect, as he once had promised in the *Mysterium cosmographicum*, but at least its imperfection is regular. The figures of the cubic marriage, the cube and the octahedron, have the tightest embrace and penetrate the outer spheres of their respective inner planets. The figures of the dodecahedral marriage, the dodecahedron and icosahedron, hold their inner planets so loosely as not quite to touch the aphelial spheres. Last, that androgyne the tetrahedron achieves balance by just touching the sphere within. Kepler promised this degree of fit in chapter 3; now he is finally ready to explain why.

In proposition 47 his arguments are—and perhaps could only be— from analogy. Here he addresses only the ideal, "if the inscription . . . had been free." (Only in the next section will he compare this ideal to the actual harmonies in the heavens.) The tetrahedron is the intermediate of the three primary figures, in location and in derivation, as he has shown in chapter 1. If nothing hinders, it should by rights fit as well on the outside, toward Jupiter, as on the inside, toward Mars. To achieve a kind of symmetry, the surrounding figures, namely the cube on the outside and the dodecahedron on the inside, should err in opposite directions; one by excess, penetrating its interior sphere, and the other by defect, failing to touch its interior sphere. Finally, the secondary companions to these figures—the octahedron

---

[86] Kepler's note at *G.W.* 6.356.4 identifies points *O* as the midpoints of the *inversa latera*.

to the cube, and the icosahedron to the dodecahedron—should fit the same way as their associated primary figures.

To determine which pair of planets should penetrate their interior spheres and which pair should fall short, Kepler refers back to the preceding axiom 46: the perfection of fit should correspond to the perfection of the figures' ratios of inscribed to circumscribed spheres. This ratio is rational, 1:3, for the tetrahedron, so the tetrahedron should fit exactly between Jupiter and Mars. The ratio of the cube and octahedron, $1:\sqrt{3}$, is irrational but rational in square. Hence those figures fit in the next-most perfect way, touching both their spheres but penetrating into the interior one. The dodecahedron and icosahedron have a ratio not even rational in square, so they fit in the least perfect way, falling short of their interior spheres.

There remains the echinus or "dodecahedral hedgehog," the regular but nonconvex polyhedron, discovered by Kepler, that fits between the spheres of Mars and Venus, straddling Earth. This figure, with its twelve interpenetrating star-shaped sides, is related to the dodecahedron and its partner the icosahedron; but Kepler points out that it is similar also to the tetrahedron. The radius of the sphere inscribed within its turned-in edges is commensurable in length to the distance between the tips of two adjacent spikes.[87] This commensurability offsets the imperfection of the proportion between inscribed and circumscribed spheres. Hence an interposition of middling quality is appropriate: not the pentagonal faces but their edges are tangent to the sphere within.

48. *Proposition.* The inscription of the regular solid figures between the planetary spheres was not done freely; for in details it was impeded by the harmonies established among the extreme motions.

This proposition summarizes chapter 9, to this point. In his discussion, Kepler promises yet more. "Let us therefore extract, from the harmonies, the intervals of the planets from the Sun, using a method of calculating that is new and never before attempted by anyone."[88]

Eccentricities From Harmony

Kepler to this point has tried to show that the world is as harmonic as it could possibly be; that the harmonies, in other words, were estab-

---

[87] According to Caspar, the radius of the sphere inscribed to the edges equals half the distance between two adjacent spikes. *G.W.* 6.553, note to 355.26.

[88] Ibid., 6.356.22–23.

TABLE 9.12.
Eccentricities Implied by Proper Proportions

1 Planet	2 Proper proportion	3 Square roots	4 Orbital radius	5 Eccentricity	6 Normed to 100,000
Saturn	$\dfrac{64}{81}$	80 90	85	5	5,882
Jupiter	$\dfrac{6,561}{8,000}$	81,000 89,444	85,222	4,222	4,954
Mars	$\dfrac{25}{36}$	50 60	55	5	9,091
Earth	$\dfrac{2,916}{3,125}$	93,531 96,825	95,178	1,647	1,730
Venus	$\dfrac{243}{250}$	9,859 10,000	99,295	705	710
Mercury	$\dfrac{5}{12}$	63,250 98,000	80,625	17,375	21,551

lished by an omnipotent Geometer. He proceeds in section 48 to complete his deduction of astronomy from harmony, by calculating the planetary orbits and motions that the optimal harmonies imply. His first step is to determine the eccentricities needed to produce each planet's proper proportion of extreme motions. The procedure is simple, as shown in table 9.12.

The proper proportions in column 2 have been established by earlier propositions in chapter 9 as the most harmonious possible, given the constraints of geometry and of other competing harmonies. Since the ratio of the apparent extreme distances for any planet is the inverse square root of the ratio of extreme motions, Kepler provides in column 3 the square roots of the terms of the proper proportions, adding extra digits of precision freely.[89] The numbers in column 3 thus

[89] In order to get the indicated numbers for the last three planets, one must triple the terms of Earth's proportion, multiply those of Venus's proportion by 40, and multiply those of Mercury's proportion by 8. Thus for Earth the square root of 3 times 2,916, or 8,748, is 93.531, and the square root of 9,375 is 96.825. Kepler, as usual, rounds the numerical results for Mercury more roughly than those for the other planets.

represent the extreme distances of each planet from the Sun, using a different arbitrary scale for each planet.

The orbital radius of a planet (column 4) is simply the average of the extreme distances; and the eccentricity (column 5) is the difference between the radius and either of the extreme distances. Both are in the scale Kepler uses to express the extreme distances for the planet. In the last column he scales the eccentricities for each planet to the size they would have if the orbital radius in column 4 equaled 100,000. This completes the task Kepler had assigned himself when he entitled chapter 9 "The Origin of Eccentricities in Individual Planets from Attention to the Harmonies among Their Motions."

Kepler has discovered the relation we know as his third law since beginning chapter 9, and presumably since giving it its title. This relation enables him to go considerably farther, to calculate not only the eccentricities but the relative sizes of the different orbits from the harmonies in the motions.

The first step is to express the mean apparent motions of all the planets in a common scale. For this he needs to express all the extreme apparent motions in a common scale, after which a theorem from chapter 3 will yield the mean motions. Table 9.13 shows the left

TABLE 9.13.
Data for Computing Mean Apparent Motions

1 Harmonic proportions of pairs				2 Extreme planetary motion	3 Proper proportion	4 Extreme motion	5 Prime factors of column 4
1				Saturn aphelial	64	139,968	$2^6.3^7$
	1			Saturn perihelial	81	177,147	3^{11}
	2			Jupiter aphelial	6,561	354,294	2.3^{11}
	5			Jupiter perihelial	8,000	432,000	$2^7.3^3.5^3$
	24			Mars aphelial	25	2,073,600	$2^{10}.3^4.5^2$
	2			Mars perihelial	36	2,985,984	$2^{12}.3^6$
32	3			Earth aphelial	2,916	4,478,976	$2^{11}.3^7$
		5		Earth perihelial	3,125	4,800,000	$2^9.3.5^5$
		5		Venus aphelial	243	7,464,960	$2^{11}.3^6.5$
1	3		8	Venus perihelial	250	7,680,000	$2^{12}.3.5^4$
		5		Mercury aphelial	5	12,800,000	$2^{12}.5^5$
4				Mercury perihelial	12	30,720,000	$2^{14}.3.5^4$

portion—the data—of Kepler's table, rearranged for clarity. The numeric values of the extreme motions, in column 4, have not appeared earlier in the book. They look like astronomical data, but in fact they are calculated from the harmonic proportions in columns 1 and 3. To make this easier to see I have added column 5, which contains the prime factors of the numbers in column 4.

The twelve numeric values Kepler uses in column 4 as the extreme motions of the planets are—quite simply—the smallest integers for which all the proportions in columns 1 and 3 hold. It is easy to verify from the factorizations that, first, the harmonic proportions are exact; and second, these are the smallest integers for which the proportions are exact. (Proof: the perihelial motion of Saturn and the aphelial motion of Mercury have no common factor.)

In this table, then, Kepler is not testing his harmonic theory but rather developing its implications. Aside from the profound investigations into planetary astronomy underlying the harmonies themselves, there is no observational component to the "data" here. The table also shows, incidentally, that the harmonic proportions among different planets listed in column 1 were complete. If Kepler had believed the proportions among any other planetary motions to be exactly harmonic, he could easily have included them in his calculations. The proportions in columns 1 and 3 of this table thus represent all the harmonies Kepler thinks he has found in the motions of the planets.

The next step is to determine the mean motions implied by the harmonic data. In order to calculate mean motions from extreme motions Kepler uses the twelfth proposition from chapter 3. This states that a planet's mean apparent motion is smaller than the geometric mean of its two extreme apparent motions by half as much as that geometric mean is smaller than the corresponding arithmetic mean. The straightforward calculations described by this convoluted proposition are shown in table 9.14.[90]

Column 3 is the proper proportion of a planet's extreme motions, repeated from table 9.13. (I have split Kepler's table, which is very wide, into two.) Columns 6 and 7 contain the arithmetic and geometric means of the two terms in each planet's proper proportion. Column 8 is half their difference, and column 9 equals column 7 minus column

---

[90] The periods in columns 6–9 of this table are decimal points, contrary to Kepler's usual practice of using a period to indicate a proportion. Kepler acknowledges the novelty with a note stating that "the figures beyond the point pertain to the precision of the number in tenth parts." *G W.* 6.357.40.

TABLE 9.14.
Calculation of Mean Apparent Motions

	3	6	7	8	9	10
Planet	Proper proportion	Arithmetic mean	Geometric mean	$\frac{1}{2}$ the difference	Mean motion	Mean motion in common dimension
Saturn	$\frac{64}{81}$	72.50	72.00	.25	71.75	156,917
Jupiter	$\frac{6,561}{8,000}$	7,280.5	7,244.9	17.8	7,227.1	390,263
Mars	$\frac{25}{36}$	30.50	30.00	.25	29.75	2,467,584
Earth	$\frac{2,916}{3,125}$	3,020.500	3,018.692	.904	3,017.788	4,635,322
Venus	$\frac{243}{250}$	246.500	246.475	.0125	246.4625	7,571,328
Mercury	$\frac{5}{12}$	8.500	7.746	.377	7.369	18,864,680

8. According to proposition 12 from chapter 3, this is the planet's mean motion. It is in the scale of the two terms in column 3 from which it was computed.

To convert these mean motions to a common scale one must use the proportionality between the terms in column 3 and the original "motions" from column 4 of table 9.13. For Saturn, the ratio of 139,968 to 64 is exactly 2,187, as is the ratio of 177,147 to 81. Multiplying Saturn's mean motion of 71.75 by this proportionality factor, one obtains the mean motion in a common dimension as 156,917.25, which Kepler rounds to 156,917. The mean motions of the other planets are obtained identically.[91]

The next step is to convert the mean motions implied by the harmonic proportions into distances from the Sun. This conversion depends entirely on Kepler's discovery that the cubes of the mean distances are proportional to the squares of the periodic times, hence inversely proportional to the squares of the mean apparent motions. Table 9.15 shows the conversion.

[91] Recalculation gives Mercury's mean motion, the last number in table 9.14, as 18,864,640 rather than 18,864,680. Caspar does not remark this; Frisch actually publishes a calculation that purports to give the incorrect result: Opera Omnia 5.487, on the last line.

TABLE 9.15.
Mean Distances from Mean Motions

1	2	3	4
		Mean motions:	Mean distances to be found among the squares
Planet	Original	Inverse, to be sought among cubes	
Saturn	156,917	29,539,960	9,556
Jupiter	390,263	11,877,400	5,206
Mars	2,467,584	1,878,483	1,523
Earth	4,635,322	1,000,000	1,000
Venus	7,571,328	612,220	721
Mercury	18,864,680	245,714	392

The mean motions in column 2 are the final result (column 10) of the previous table. The inverse mean motions in column 3 can be obtained by dividing the numbers in column 2 into Earth's mean motion of 4,635,322 and multiplying by 1,000,000. Christopher Clavius had conveniently published a table containing both squares and cubes,[92] so Kepler looks up the mean motions among the cubes and noted the corresponding entry among the squares. He thus obtains numbers proportional to the mean distances, in a common dimension. (Column 4, in other words, is proportional to the $\frac{2}{3}$ power of column 3.) The distances Kepler publishes, column 4 in table 9.15, have been divided by 10 in order to norm Earth's mean distance to 1,000.

A previous table (table 9.12) has given the eccentricities implied by Kepler's harmonies. Converting these to the common dimension, he produces the aphelial and perihelial distances in that same dimension. Table 9.16 shows the results.[93]

---

[92] C. Clavius, *Geometriae practicae* (Rome, 1604), Kepler had received this book around the beginning of October 1606 from one J. R. Ziegler in Mainz; see letters 392 and 395 in *G.W.* 15.341, 353. At the end of book 8, Clavius printed a table with columns of *Radices* (from 1 to 1,000), *Quadrati*, and *Cubi*.

[93] Mercury's extreme distances in this table obviously differ by 84 from the mean

Table 9.16.
Orbital Dimensions Implied by the Harmonies

1	2	3	4	5
Planet	Mean distance	Eccentricty	Aphelial distance	Perihelial distance
Saturn	9,556	562	10,118	8,994
Jupiter	5,206	258	5,464	4,948
Mars	1,523	138	1,661	1,384
Earth	1,000	17	1,017	983
Venus	721	5	726	716
Mercury	392	85	476	308

EMPIRICAL ACCURACY

Kepler has calculated the distances and eccentricities in table 9.16 based on his harmonic law and the six proper proportions and eight paired proportions in columns 1 and 3 of table 9.13. These distances and eccentricities summarize the observable consequences of the harmonic structure developed in book 5. One might reasonably expect, at last, a detailed comparison of them with the observed distances and eccentricities; but Kepler only remarks that "all approach very closely to those intervals which I found from the Brahean observations. In Mercury alone there is a small difference."[94]

The numbers are at hand; let us compare them ourselves. Kepler's observed distances, from chapter 4, are in table 9.17.

The two sets of numbers are indeed close; it is clear why Kepler feels no great need to apologize. He ascribes the largest discrepancy, that

---

distance, although Kepler, Frisch, and Caspar all print 85 for Mercury's eccentricity. Calculating forward with the third law from Kepler's reported mean distance of 392 indeed gives 84.48, but calculating from a more exactly calculated mean distance of 392.3 gives 84.54. The numbers for Mars contain equally trivial rounding error.

[94] G.W. 6.359.2–4.

TABLE 9.17.
Observed Orbital Dimensions

| 1 | 2 | 3 | 4 |
| | Mean | Aphelial | Perihelial |
Planet	distance	distance	distance
Saturn	9,510	10,052	8,968
Jupiter	5,200	5,451	4,949
Mars	1,524	1,665	1,382
Earth	1,000	1,018	982
Venus	724	729	719
Mercury	388	470	307

for Mercury, either to lack of sufficient observations to calculate its orbit well or to its large eccentricity.[95]

The latter possibility—that Mercury's disparity is due to the size of its eccentricity—is a recurrent theme in the *Harmonice mundi*. Kepler refers his doubts here only to chapter 3, which could aim at either of two places in that chapter. The first is proposition 6, which states that the apparent diurnal arcs are approximately proportional to the inverse square of distance from the Sun, *provided that the eccentricity is not very large*. Kepler has used this proposition explicitly in the discussion leading up to table 9.12, where he calculates the extreme distances by taking the square roots of the corresponding diurnal arcs. In doing so, he was considering the apsidal motions, however, and at the apsides the inverse-square proportionality does not depend upon the eccentricity being small. Kepler surely knew this.

The other small-eccentricity approximation acknowledged in chapter 3 is proposition 8, the original statement of the $\frac{3}{2}$-power rule. Kepler qualified this rule by requiring that the mean of the orbit's two diameters be not much less than the longer diameter, an oblique way of insisting that the eccentricity be small. The discrepancy between the observed and predicted sizes of Mercury's orbit might well be one

[95] Ibid., 6.359.5–7.

source of his uncertainty regarding that recently discovered rule. In 1618 Kepler had no very strong reason to prefer a recent inspiration such as the period-distance relation to his painstakingly perfected harmonic edifice.

HARMONIES AND POLYHEDRA

The final table in chapter 9, and indeed in the *Harmonice mundi*, is a comparison between the dimensions derived from celestial harmonies and those implied by the regular polyhedra. As with many of Kepler's tables, the rows of table 9.18 represent so many statements, here eight, with the column headings supplying the framework needed to form the words and numbers in each row into sentences. The first row yields: "If the radius of the sphere circumscribed to the figure, which is ordinarily taken to be 100,000, becomes, in the *cube*, *8,994*, then the radius of the sphere inscribed from *Saturn*, instead of *57,735*, becomes *5,194*. Whereas the *middle* distance of *Jupiter* arising from the harmonies is *5,206*." The table, in other words, applies a polyhedron's proportion of inscribed to circumscribed spheres to a planetary sphere, to obtain the size of the next sphere inscribed to the polyhedron. It then compares this to the size determined harmonically by the earlier propositions in chapter 9.

The radii in column 2 are (with one exception in the last row) the perihelial distances of the corresponding planet from column 5 of table 9.16, because the figures are nested within the innermost or perihelial sphere of the planet outside them. These distances are calculated from harmonic proportions and the period-distance relation, as we have seen. They are an appropriate starting point for the application of polyhedral proportions, since the purpose here is to compare the distances resulting from the harmonic theory with those from the polyhedral theory.

The bottom two rows of table 9.18 attempt (unsuccessfully) to use the proportions of the "octahedron's square," the square connecting the four points around the waist of an octahedron. Kepler had suggested in the *Mysterium cosmographicum* that Mercury's aphelial sphere was perhaps inscribed within this square rather than within the octahedron itself. Here he shows that the proportion of the octahedral square is too small to fit well between the perihelial sphere of Venus and the aphelial sphere of Mercury, and also (on the last row of table 9.18) too small to fit between the aphelial and perihelial spheres of Mercury. He is perfectly willing to discard this idea of

TABLE 9.18.
Harmonies and Polyhedra

1	2	3	4	5	6	7	8
If the radius of the sphere circumscribed to the figure, which is ordinarily taken to be 100,000, becomes, in the		*then the radius of the sphere inscribed from*	*instead of*	*becomes*	*Whereas the*	*distance of*	*arising from the harmonies is*
cube,	8,994,	Saturn,	57,735	5,194.	middle	Jupiter	5,206.
tetrahedron,	4,948,	Jupiter,	33,333	1,649.	aphelial	Mars	1,661.
dodecahedron,	1,384,	Mars,	79,465	1,100.	aphelial	Earth	1,018.
icosahedron,	983,	Earth,	79,465	781.	aphelial	Venus	726.
hedgehog,	1,384,	Mars,	52,573	728.	aphelial	Venus	726.
octahedron,	716,	Venus,	57,735	413.	middle	Mercury	392.
octahedral square,	716,	Venus,	70,711	506.	aphelial	Mercury	476.
or	476,	Mercury,	70,711	336.	perihelial	Mercury	308.

his youth; after all, "what business do plane figures have among solid?"[96]

The numbers in column 4 are the polyhedral proportions of inscribed to circumscribed spheres. All except the octahedron's square were given in chapter 1 of book 5. Indeed all except the hedgehog had been given, to three digits' precision, in chapter 14 of the *Mysterium cosmographicum*. Here they are scaled to a circumscribing sphere of radius 100,000. Upon being multiplied by the perihelial distances in column 2 (also taken as a proportion of 100,000) they yield the numbers in column 5 as predicted radii of the inner sphere. Columns 6 and 7 identify which sphere of the inner planet agrees most closely with the result in column 5.

The point of the table is to compare column 8, the harmonic radii from column 2, 4, or 5 of table 9.16, with the polyhedral radii in column 5. The two sets of radii agree well although by no means perfectly. The agreement is not surprising; after all, the polyhedra supplied one of the axioms from which Kepler deduced the harmonies. He recites the discrepancies between the models, in an affectionate apology for his *Mysterium cosmographicum*. The nested polyhedra had served the Creator, as they served Kepler, as a rough model, fixing the number and approximate dimensions of the planetary spheres. Harmonic proportions among the apparent motions of the planets then determined the final dimensions.

A FINAL PUZZLE

Kepler briefly considers one last adjustment to the harmonic structure. If that small increment or *additamentum*, equal to the proper proportion of Venus, by which he increased the proper proportion of Mercury in proposition 45, had instead accrued to the converging proportion of Venus and Mercury, then the dimensions predicted for Mercury would "be very precisely represented by Astronomy."[97] Obscure possibilities like this were of real interest to Kepler, as he tried to maximize both the harmonic beauty of his structure and its empirical accuracy. The proposed adjustment is shown in table 9.19.

Kepler's unmodified theory put the three innermost motions in the eminently harmonic proportions 3:5:12. Reducing the proper proportion of Mercury by 243:250, and increasing the converging proportion

---

[96] Ibid., 6.359.26.
[97] Ibid., 6.359.36–38.

TABLE 9.19.
Suggested Correction to Mercury

1	2	3	4
		Orginal theory in lowest & comparable	*Alternate*
Extreme motion		*terms*	*theory*
Venus perihelial	3	729	729
Mercury aphelial	5	1,215	1,250
Mercury perihelial	12	2,916	2,916

of Venus and Mercury by the same amount, yielded the much less pleasing proportions 729:1,250:2,916. The cosmological effect was to increase the aphelial motion of Mercury.

One can recalculate the last row of table 9.14 to see the effect of this adjustment on the mean motion and distances of Mercury. In table 9.20 Mercury's proper proportion has been hypothetically reduced to 1,250:2,916, or 625:1,458, as in column 4 above. In converting the mean motion into a common dimension for column 7, one must use extreme motions of 13,168,724 (rather than 12,800,000) and 30,720,000 since the converging proportion of Venus and Mercury has increased to 729:1,250.

By this calculation the mean Mercurial motion would be nearly 2 percent faster than it was in table 9.14. The period implied by this motion, with the $\frac{3}{2}$ power rule, yields a mean distance of 388 (table 9.21), as Kepler states.

Since the eccentricity implied by Mercury's proper proportion of 625:1,458 was now 20,866 (normed to a radius of 100,000), the new extreme distances were 469 and 307. As Kepler points out, these distances agree very precisely with observation.[98]

He rejects the modification nonetheless. It disrupts Mercury's proper harmony and its joint harmonies with Venus; it removes Mercury's aphelial motion from any musical scale; and if that were not

[98] Ibid., 6.359.36–38. The adjustment affects only the aphelial motion and distance of Mercury. The ostensible reduction of the perihelial distance from 308 to 307 arises from the rounding of intermediate results.

TABLE 9.20.
Revised Calculation for Mercury

Planet	5 Proper proportion	6 Arithmetic mean	7 Geometric mean	8 $\frac{1}{2}$ difference	9 Mean motion	10 Common dimension
Mercury	$\frac{625}{1.458}$	1,041.5	954.6	43.45	911.15	19,197,893

enough, it spoils the analogy between the harmonies of the highest and lowest planets.

Kepler also cites a further reason, seemingly more objective, for rejecting the change: that the revised distances would increase the planet's mean diurnal motion and hence diminish its periodic time. Mercury's periodic time, however, is known very accurately from centuries of observation. Kepler calls it "the most certain of all astronomy,"[99] by which he evidently meant that more revolutions of Mercury had been counted, over the centuries, than of any other planet. Hence the division of revolutions into time should yield a more accurate result for the period of Mercury than for any other planet.

This second, properly astronomical reason presents a puzzle, for the adjusted mean distance of 388 agrees exactly with the distance Kepler has deduced from Mercury's periodic time (shown in table 9.17). Adjusting Mercury's aphelial motion by the *additamentum* actually improves the agreement between the observed period and the period calculated from the harmonies. I cannot explain Kepler's assertion to the contrary. His real reasons for rejecting the change in Mercury's aphelial motion surely arise from its disruption of the harmonies and not from the alleged inaccuracy in the periodic time it implies for Mercury.

Summary

Kepler calls the forty-ninth and last section of chapter 9 neither an axiom nor a proposition but rather an *epiphonema*, or summary.

49. *Summary.* It was right that in the origin of the intervals among planets, the solid figures yielded to harmonic reasons, and the greater

[99] Ibid., 6.360.2–3.

TABLE 9.21.
Revised Orbital Dimensions for Mercury

1	2	3	4
		Mean motions:	Mean distances to be found among the squares
Planet	Original	Inverse, to be sought among cubes	
Mercury	19,197,893	241,450	388

harmonies of two planets yielded to the universal harmonies of all, to the extent that this was necessary.

It is a pleasant coincidence, Kepler remarks, that the development of his thesis has reached section 49, the seventh square number, after six preceding octuples of axioms and propositions in the first forty-eight sections. He could include this particular assertion among the axioms, he writes, but has preferred to call it a summary "since God also, having completed the work of creation, looked upon all the things that He had made, and lo! they were very good."[100] Clearly Kepler, in proposing such a presumptuous analogy, is mightily pleased, himself, with the work he has completed.

Despite the axiomatic character of what is asserted here, Kepler discusses or "demonstrates" each of its two halves. The first half, that harmonic reasons took precedence over the regular polyhedra, calls forth from Kepler's pen an extended metaphor comparing Geometry to Body and Harmonies to Form. To start with, the solid figures are geometrical, and intimately linked to the static spaces between the planets' spheres and the number of those spaces; hence they determine the number of planets. The harmonies he has discovered, on the other hand, are relations among motions of the planets. Form is what gives precise boundaries to matter and limits it to a definite shape. Among the infinite number of proportions found in geometry, only a few are harmonic. These harmonic proportions arise from a finite class of especially perfect plane figures. They complete the geometry of the figures, as a sculptor's shaping completes the stone:

[100] Ibid., 6.360.21–25.

As therefore matter desires form, as a rough-cut stone of the right size desires the shape of the human body, so the proportions of the geometric figures desire harmonies: not that the latter fashion and shape the former, but because certain matter agrees more aptly with a particular form, a certain quantity of stone with a particular likeness, a certain figural proportion with a particular harmony; in order that they may be fashioned and shaped more completely, the matter by its form, the stone by a sculptor into the image of life, the proportion of the figural spheres by its neighboring and appropriate harmony.[101]

Kepler clarifies this metaphor—at least he says that he is clarifying it—with a brief account of the stages by which he arrived at his harmonic theories. Some twenty-four years earlier—hence in 1595 or perhaps 1594—he began contemplating such matters.[102] He wondered then whether the planetary spheres were equally distant from one another, since equality was the most beautiful of proportions. That idea, he has since realized, lacks both head and tail, since there is no limit at either end to the number of spheres that can be constructed in fixed proportion to one another. This is a problem that afflicts many other schemes of proportionality: a theory that accounts for an infinite number of planetary spheres is unsatisfactory as an explanation for six of them. Regular polygons, for example, might seem expressly designed to account for the spacing between the spheres, but they cannot explain how many spheres there are.

The five regular polyhedra solved this problem for Kepler and did reasonably well in accounting for the spacing between the spheres of the six planets. Quite reasonably, he had awaited "the perfection of Astronomy" for the resolution of the spacing errors. Astronomy in 1596, as Kepler realized perhaps better than anyone, was in the midst of a thorough reconstruction. By 1618 the work was done; he had carried it out himself. Still the intervals disagreed with the polyhedral figures; nor was any explanation in sight why the eccentricities were unequally distributed among the planets: "I had, in this house of the World, been looking for nothing but the stones, of a more elegant shape perhaps, but still one suited for stones; not realizing that the

---

[101] Ibid., 6.361.8–15.

[102] The account that follows parallels that in the 1596 preface to the *Mysterium cosmographicum*. The earliest surviving sketches of Kepler's polyhedral theory are in three letters to Maestlin dated August, September, and October of 1595, letters 21, 22, and 23, in *G.W.* 13.27–46.

Architect had shaped it into a very detailed effigy of a living body. And so little by little, especially these last three years, I arrived at Harmonies, leaving behind the solid figures in matters of detail."[103] Harmonies took over from the polyhedra both because they gave a more detailed shape, a Form, to the rough picture provided by the Matter of the solid figures and because they depended upon the eccentricities of individual planets, and hence constrained those eccentricities to be exactly what they were observed to be. The harmonies, Kepler wrote, attached the nose, the eyes, and the other members to the statue whose rough proportions were given by the polyhedra.

The second part of the summary making up section 49 asserted that universal harmonies of all the planets had taken precedence over harmonies of two planets—in particular, over the "greater" harmonies of two planets, which is to say the harmonies of their diverging motions. This is not surprising; the more planets are united in a harmony, the greater that harmony is. When one harmony must yield to another, because it is mathematically impossible for both to hold exactly, the lesser harmony is the one that yields. Among harmonies of two planets, Kepler continues, the harmonies of diverging motions yield to those of converging motions because diverging motions "look at" other planets, whereas converging motions look at each other. (Harmonies between motions that look at additional planets apparently do not receive the same credit for universality as harmonies that actually involve the motions of additional planets.)

Having demonstrated that the order of precedence observed among the pleasing proportions in the heavens is as it should be, Kepler closes this extended ninth chapter with a prayer. He squeezes the whole prayer, as usual for him, into a single sentence, which we join in progress:

> . . . O You who by the light of nature arouse in us a longing for the light of grace, so that by means of that You can transport us into the light of glory: I give thanks to You, Lord Creator, because You have lured me into the enjoyment of Your work, and I have exulted in the works of Your hands: behold, now I have consummated the work to which I pledged myself, using all the abilities that You gave to me; I have shown the glory of Your works to men, and those demonstrations to readers, so far as the meanness of my mind can capture the infinity of it, for my mind was

---

[103] *G.W.* 6.361.32–36.

prepared for the most perfect philosophizing; if anything unworthy of Your deliberations has been proposed by me, a worm, born and raised in a hog-wallow of sin, which You want mankind to know about, inspire me as well to change it; if I have been drawn by the admirable beauty of Your works into indiscretion, or if I have pursued my own glory among men while engaged in a work intended for Your glory, be merciful, be compassionate, and forgive . . . .[104]

Chapter 10: Inhabitants of the Sun

In the final chapter of the *Harmonice mundi* Kepler turns his attention, in his words, "from the celestial music to the listener."[105] He puts aside the question of *what* harmonies are in the heavens to explain *why* harmonies are in the heavens. To what purpose, he asks, has the intricate polyphony among the motions been established, the precisely calculated modulations between durus and mollis harmony, the scrupulous attention to reducing even the small dissonances necessarily entailed by those modulations?

He has found harmony by comparing the angular velocities of the planets. There is nothing harmonious in the individual motions or indeed in anything that can be perceived directly from the planets individually. The harmonies are rather in the comparison of different motions as they appear from the Sun. Their purpose is somehow connected with the Sun, therefore. The planetary harmonies—indeed any harmonies—involve proportions, and hence a comparison of two or more terms. This is an important clue, for comparison is an activity that requires a mind. "Of course," Kepler admits, "it is not easy for us, dwelling on Earth, to conjecture what kind of vision might be in the Sun, what eyes, or what other instinct for perceiving these angles even without eyes, and for estimating the harmonies of the motions entering the vestibule of a Mind by some gate; what, finally, that Mind might be, in the Sun."[106]

Not only is the Sun the source of light and heat for the Earth and the other planets, but conversely from all regions of the world harmonies are returned to the Sun, as if repaying the debt. Time after time, the appearances of two motions come together and combine to form a

[104] Ibid., 6.362.40–363.13.
[105] Ibid., 6.363.16.
[106] Ibid., 6.363.28–32.

single harmony. The synthesis of a harmony from two sense percep-
tions can only take place in a Mind. The Sun, as the source of the force
that moves the planets, is the seat of government of the kingdom of
nature, Kepler reflects; surely it is populated with suitable chancellors,
governors, or prefects.

These considerations induce Kepler to speculate on spiritual beings,
such as might be thought to reside on or in the Sun. He quotes liberally
from the hymns of Proclus, whose neo-Platonism is agreeable to such
speculation. Proclus was not a Christian, of course, and Kepler is aware
of the danger in associating himself too closely with pagan ideas on
such matters. He emphasizes prudently that he himself does not believe in "intelligent Gods," like Aristotle, or in innumerable multitudes
of planetary spirits, like those he calls "the magicians." With these
cautions on record, he wants in chapter 10 to inquire what sort of
Mind, or what kind of intelligent beings, could inhabit the Sun. Following a thread of analogy, as he says, he wants to penetrate the labyrinths
of the mysteries of Nature.

The analogy Kepler follows is this. The relation between the planets
circulating about the Sun and the Sun itself is like the relation of
discursive thought to simple mental reflection. The Sun does not
move, but rotates in place, as is shown by the motions of sunspots
across its face. By means of this rotation it somehow moves the planets
around it. Analogously, the simplicity of mind itself moves the discursive thoughts that analyze the world and make it understandable. If
the Earth, our domicile, did not travel around the Sun, human reasoning would never have learned the true intervals between the planets,
because it is the observer's motion that permits him to judge the
changing distances of the planets from one another and from the Sun.
Without the motion of the Earth, the construction of true astronomy
would be impossible. For intelligences on the Sun, however, only the
angular motions (and the harmonies latent in them) can be perceived.
No knowledge of distances to the planets is possible—unless, of
course, that knowledge has been furnished a priori by the Creator.

To be sure, it is an unusual concert that the residents of the Sun
attend. Their attention is fixed on very slow motions of very distant
bodies. On average Saturn (seen from the Sun) takes two weeks to
move through an arc equal to the width of the full Moon (seen from
Earth). The solar intelligences not only must perceive this motion,
they must perceive—and enjoy—the moments when Jupiter is moving exactly twice as fast as Saturn. In these delectable moments, Jupi-

ter moves fast enough to rush across a Moon's-width of sky in about a week. The awareness of the spectators is such that they can savor the comparison of such motions. The only alternative to an extremely slow perception of time is an equally extreme acuteness of vision, or whatever passes for vision among them. (If Saturn's position could be perceived to fourth minutes of arc, say, then its mean motion of some 5$'''$ per second might be quite noticeable, and easily comparable to Jupiter's occasional runs of 10$''''$ per second.) This kind of visual acuity seems unlikely in view of the extremely wide field of vision needed to track all six planets; but what do we know? Kepler thinks the beings in the Sun are simple; he must have also thought them rather slow.

Having gone this far, and feeling "warmed by having drunk a generous draught . . . from the cup of Pythagoras which Proclus offers," Kepler proposes to let himself doze off listening to the sweet harmonies of the planetary choir, and dream about the beings that might inhabit the planets themselves.[107] These would be of a different kind from those in the Sun. Intellect in the Sun is pure and simple, the source of all harmony in the world. Powers of discourse and reasoning pertain to the globes swept around the Sun, and in particular to Earth and its inhabitants. Beings on the other planets would therefore resemble us much more than they would resemble the pure and simple beings on the Sun. Like us, residents of other planets would have no direct apprehension of the harmonies, but like us they would be able to attain a knowledge of astronomy, and perhaps eventually of the planetary harmonies, by observing and reasoning.

His enthusiasm for the subject growing, Kepler dreams in expansive detail about what a waste it would have been if the huge planetary bodies had been created without their own inhabitants.

> Should we not, from the variety of the works and plans of God which we detect on this globe the Earth, arrive also at the same conjecture concerning the other globes? For he who created the forms of life that inhabit the waters, despite the absence under water of the air which living things breathe; who discharged into the vastness of the air birds, supported by wings; who gives to the snowy northern regions white bears, white foxes; and their food, for the bears the whale of the ocean, for the foxes birds' eggs; who gives to the scorching deserts of Libya lions, and to the broad plains of Syria camels, and gives a tolerance of hunger to the lions, and of thirst to the camels; did he consume all of his

[107] Ibid., 6.367.11–14.

skill in the globe of Earth, so that he was not able to adorn the other globes with suitable creatures, or did he consume all of his goodness, so that he did not want to do so?[108]

Even the solid figures embracing the Earth's orbit reflect the forms of life on the planet. As Kepler explained in the first chapter of book 5, the dodecahedron between Earth and Mars is masculine, the icosahedron between Earth and Venus feminine. The ineffable proportion shared by the inner and outer spheres of those two figures represents the joining of masculine and feminine in the act of generation. What, he asks with some curiosity, should we think of the other globes on the basis of their own geometric figures?

All the planets surely have their own inhabitants. Otherwise the common features that have been recently revealed by Galileo's telescope would have no purpose.

> For what reason do four moons accompany Jupiter with their orbits, and two Saturn, as this our only Moon follows our domicile? In the same way we will consider also the globe of the Sun. . . . Is that globe empty, and the others full, if all other things correspond more closely? If just as the Earth exhales clouds, so the sun exhales blackened soot? If just as the Earth is moistened and made green by rain showers, so the Sun ignites with those burned spots it has, brighter flames breaking out from the body which is itself afire? What is the use of this apparatus, if the globe is empty? Do not the senses themselves cry out that fiery bodies live here, capable of simple Minds?[109]

Awakening from his dream, Kepler closes the chapter with a prayer that God might be praised by the heavens, by the Sun, Moon, and planets, by the celestial harmonies and their beholders, and finally by his own soul.[110]

Angels in the Sun?

The speculations in the last chapter of the *Harmonice mundi* about beings inhabiting the Sun and other planets—combined as they were

---

[108] Ibid., 6.367.22–32

[109] Ibid., 6.367.40–368.10.

[110] Ibid., 6.368.14–22. A marginal note singled out one beholder of the harmonies by name: "And you above all, happy old Maestlin, for you used to inspire these things I have said, and you nourished them with hope."

with a disavowal of any religious implications—evidently surprised Kepler's contemporaries less than they do us. The possibility of such beings arose also in correspondence that Kepler conducted with one Vincenzo Bianchi, a nobleman and scholar residing in Venice for whom Kepler drew up a horoscope. In November of 1618 Kepler sent Bianchi the first page of book 5, fresh from the press: the *prooemium*, or preface, containing Kepler's rapturous announcement of the period-distance relation. The anxious author wanted to know about the prospects for selling his new book, or at least portions of it, in Italy; he feared that the German booksellers would set a price so high that it would find few buyers—a problem even then. He asked if a book could be sold in Italy without the *imprimatur* of the Roman Church, which his openly Copernican work lacked.[111] (The *Harmonice mundi* did not explicitly defend the motion of the Earth, but it certainly assumed a heliocentric cosmology; and Kepler had published other books that argued forcefully enough in favor of Copernicus.)

Bianchi showed the preface not only to booksellers but to a number of those whom he described as the most learned men in Venice. All of them, he reported, were thrown immediately into a state of incredible curiosity to see the remainder. Kepler's books were very difficult to obtain in Italy, and Bianchi thought that a thousand copies of the new one could be sold there at a high price. The Copernican system had been proscribed recently, "but the books of illustrious German authors, whenever they are condemned, are nevertheless sold secretly and read the more intently."[112]

The issue of Angels, whose existence Kepler had cited in book 4 as biblically attested, arose naturally in the context of the Church's attitude to the book. In a letter of February 1619, Kepler directed Bianchi's attention to this final chapter of book 5, explaining what he meant quite clearly. The harmonies among planetary motions appear only at the Sun; hence on the Sun there reside "either a God or minds such as those of a creature of the human kind on Earth." The former alternative is contrary to faith, since God is omnipresent and has no particular domicile.[113] Therefore some rational creature resides in the globe of

[111] Kepler to Vincenzo Bianchi, 30 November 1618, no. 810 in ibid., 17.288–91.

[112] Vincenzo Bianchi to Kepler, 20 January 1619, no. 825 in ibid., 17.318–19; and 6.282–83.

[113] The question arises whether an omnipresent God might not have ordained harmonies that were "apparent" from the Sun simply because the Sun is the most prominent place that there is. Kepler did not address this question; perhaps he liked to believe that the Sun had inhabitants.

the Sun, a simpler, purer intellect than the human. As in the *Harmonice mundi*, Kepler describes the solar intellects as simple because although they contemplate the planetary motions they lack any means of determining planetary distances and hence can never learn about astronomy. (Astronomy ranks high, in Kepler's estimation, as an exercise for the intellect!) He would call the sun-dwellers angels, he writes, except that angels have no particular dwelling place, since they must serve as heavenly diplomats, constantly sent on missions by God. The minds that dwell on, or in, the Sun have a very important reason for staying there: their function is to enjoy the planetary harmonies: "It does not suffice to say that those harmonies are for the sake of Kepler, and those after him who will read his book. Nor indeed are aspects of planets on Earth for the sake of astronomers, but they insinuate themselves generally to all, even peasants, by a hidden instinct. For such is the power of a birth horoscope. Meanwhile I am very careful not to introduce intelligent gods or to support magical superstitions.[114]

[114] Kepler to Vincenzo Bianchi, 17 February, letter no. 827 in *G.W.* 17.326.213–19.

Conclusions

KEPLER PROBABLY HOPED that the *Harmonice mundi* would endure longer than any of his other books. The science of harmony represented for Kepler, as it had for Ptolemy, a synthesis of all that could be known about order and beauty and about their embodiment in the world. Kepler recognized, however, that few readers would be able to appreciate the intricate harmonies he had detected in the heavens. More than once he admitted as much, as in concluding the preface to book 5: "Let it wait a hundred years for its reader, as God himself has awaited a witness for six thousand years."[1]

His pessimism has turned out to be justified. By taking the tradition of *musica mundana* very seriously indeed, and developing it to its highest level of sophistication, just at the dawn of the modern age of astronomy, Kepler exhausted the possibilities for carrying on that tradition. The response to his book was not as uniformly bad as Kepler feared; he was, after all, the imperial mathematician, a man of versatile and widely acknowledged brilliance. He was admired even when he was not understood. With few exceptions, however, he was not understood. Kepler's work was far too original to be absorbed readily into the mainstream even of astronomy, and that science itself was too highly developed to be diverted far from its course by his novel theories. This was all the more so because the astronomy in his theories was embedded in a matrix of physical speculation, archetypal reasoning, and harmonic rhapsody that other scientists found nearly impenetrable. It must have seemed that there was little hope that anyone capable of understanding the *Harmonice mundi* would take it seriously enough to appreciate the great plan whose details it traced.

RECEPTION OF THE *HARMONICE MUNDI*

By the mid–seventeenth century, readers had succeeded in plucking the three "laws of planetary motion" out of Kepler's books. They were

[1] *G.W.* 6.290.8–9.

much more likely, however, to refer to Kepler's textbook, the *Epitome astronomiae Copernicanae*, and to the *Rudolphine Tables* for such results than to the original sources, the *Astronomia nova* and the *Harmonice mundi*.[2] Kepler's theories of celestial harmony, on the other hand, seem to have been largely ignored. Soon after birth they fell into a kind of limbo, from which they have never emerged: too technical to be read by those who listen for the music of the spheres, and too peculiar to be taken seriously by scientists with the technical ability to understand them.

Two very different writers did describe the musical harmonies of Kepler's astronomy in sufficient detail to indicate that they had read the *Harmonice mundi* attentively. One was Jeremiah Horrocks, an Englishman, perhaps the astronomer who best deserves to be considered Kepler's scientific successor. The other was an Italian Jesuit named Giovanni Battista Riccioli, a widely read amateur historian of astronomy whose immense *Almagestum Novum* surveyed virtually the entire history and literature of astronomy in the middle of the seventeenth century.

Horrocks

In Jeremiah Horrocks (1618–41), Kepler found posthumously a more understanding and sympathetic reader than he ever knew during his life. Horrocks might have contributed greatly to astronomy had he not died so young. As it was, his principal work was in lunar theory.[3] An early advocate of the astronomical tables of Lansbergius, Horrocks in the 1630s transferred his allegiance to Kepler's *Rudolphine Tables*, and through study of the principles underlying them acquired a profound, and occasionally critical, knowledge of Kepler's astronomy. His *Astronomia Kepleriana defensa & promota* was published only in 1673, in a collected volume of posthumous works.

As his title suggests, Horrocks accepted Kepler's work, including most of the physical and harmonic theories, as basically sound. The book is largely about lunar theory, but in the prolegomena Horrocks

[2] Russell, "Kepler's Laws of Planetary Motion: 1609–1666," *British Journal for the History of Science* 2 (1964): 1–24. The general conclusions of Russell's survey have been refined in many respects by the work of C. Wilson, collected in *Astronomy from Kepler to Newton* (London, 1989).

[3] See C. Wilson, "On the Origin of Horrocks's Lunar Theory," reprinted in his collection *Astronomy from Kepler to Newton*.

addressed larger questions. Regarding the harmonic speculations in particular, he observed that Kepler "first (and thus far alone) has sailed this sea." Horrocks recognized that the *Harmonice mundi* went well beyond the speculative writings in the *musica mundana* tradition. The results, he thought, were remarkably good for a first attempt at such an ambitious project. In particular he was impressed that Kepler had used harmonic reasoning to demonstrate the proportions among the heavenly spheres "to the minute" of arc.[4] This was probably a reference to the harmonic law relating the planets' well-known periodic times to their distances from the Sun.

Horrocks also admired the derivation of the planets' eccentricities from specifically musical speculations. Admitting that the calculated values were not as accurate as one might desire, he was still willing to excuse the error. "It could be believed that the harmonic eccentricities, exquisitely constituted in the beginning, now by the long passage of time have been corrupted, and by accidental physical causes changed somewhat from their original measure."[5]

Indeed Horrocks was fully aware of the complexity introduced into astronomy by Kepler's physical theories. He cited approvingly Kepler's intricate calculation, based on hypothetical physical causes, to explain why a year contained $365\frac{1}{4}$ days rather than the archetypal number 360. Going even beyond Kepler, Horrocks tried to unite physics and harmonics, suggesting that some of the harmonic proportions in the heavens originated in a kind of physical resonance. When one string is plucked, it can make a nearby string vibrate as well. The same thing, he proposed, takes place in the celestial harmony. The Sun might be thought to transmit its forces to the Earth, as one string to another, giving rise to a chord in a celestial music formed by the motions of the Earth around the Sun and the Moon around the Earth. This was an attempt to introduce the physical theory of music, as observed in vibrating strings, into the physics of planetary motion. It was not a Keplerian theory—Horrocks thought that harmonies of this kind would not occur among the different primary planets—but it was obviously inspired by both the physical and the harmonic aspects of Kepler's planetary theory.

Horrocks died in his early twenties and so had no time to try to develop any of these ideas further. If he had, one suspects that he could not have taken them much farther than Kepler did.

---

[4] J. Horrocks, *Opera posthuma* (London, 1673), p. 14.

[5] Ibid., p. 11.

Riccioli

Giovanni Battista Riccioli (1598–1671), by vocation a Jesuit and teacher, collected astronomical theories as an avocation. He exhibited the range of his learning in an ambitious work, never quite finished, entitled *Almagestum Novum*.[6] The work was intended to be divided into two tomes, of which the second never appeared—quite excusably, given the scale of the first, which itself occupies two massive volumes. Book 9 of tome 1 is devoted to systems of the world, and is divided into five sections: first, "The Creation and Nature of the Heavenly Bodies"; second, "The Motions and Movers of the Heavenly Bodies"; third, "Systems of the World Supposing that the Earth Does Not Move"; fourth, "The System Supposing the Earth to Move"; and fifth, "The Harmonic System of the World." Each section is further divided, of course, into a suitable number of chapters, for the author has a great deal of material to cover.

Riccioli professes an orthodox belief that the Earth does not move. He shows a distinctly open mind, however, in describing the merits of all points of view, to the point where one inevitably begins to wonder where his sympathies lie. The world of a Jesuit scientist in the seventeenth century contained many subtleties, however, and I am not prepared to speculate on what Riccioli really thought. Certainly in the fourth section of book 9, while describing the heliocentric system of the world, he gives a thorough presentation of Kepler's astronomy, physics and all. He apparently regards Kepler's version of the heliocentric system as the most accurate and technically sophisticated hypothesis available, although he declares it unacceptable on philosophical and religious grounds.

In the fifth section Riccioli begins by listing all the writers he knows—this includes a lot of writers—who have treated the harmony of the heavens. The references cascade down the page, making the *Almagestum Novum* an essential bibliographic tool in this, as in all other areas relating to Renaissance astronomy. Taking up in rapid succession a miscellany of fundamental questions, such as whether the heavenly bodies make a real sound by their motion, and what proportions are rightly regarded as consonances, Riccioli cites all the opinions he can remember or lay his hands on. The theories of Kepler appear prominently, and are treated always with respect.

Finally in the eighth chapter of the fifth section of book 9 of tome 1,

---

[6] G. B. Riccioli, *Almagestum Novum* (Bologna, 1651).

Riccioli takes up the proposals of various authorities to establish the distances to the planets on harmonic principles. He makes it quite clear that he is evaluating these theories in terms not of elegance but rather of scientific truth. Summarizing the theory of the Pythagoreans, he pronounces it *absurdissima*. He then moves on to Pliny (and Censorinus, Zarlino, Giorgio Valla, and Glareanus, all of whom Riccioli summarily groups together) and the Platonists (Macrobius and Marsilio Ficino), all of whose opinions he likewise judges absurd. Then he turns to Mario Bettini, a writer whom we have not yet encountered.

Bettini (1582–1657), also an Italian Jesuit, was in 1651 in the process of writing a treatise entitled *Apiariae Universae Philosophiae...*, published piecemeal, in quite a few parts, between 1641 and 1660. Since Bettini's work is virtually unknown today, I reproduce in table 10.1 the distances Riccioli reports from the Earth to the various spheres of Bettini's geocentric world, and the associated musical intervals.[7]

I have not seen Bettini's work itself, so I do not know the extent, if any, to which it was consciously written as a geocentric alternative to Kepler. The distances in column 3 of the table are peculiar, to say the least, in the mid–seventeenth century. The numbers are evidently given in terrestrial radii, since the solar distance of 1,145 is close to all the standard geocentric estimates of that distance in terrestrial radii.[8] Riccioli reports that Bettini is using the proportions treated as consonances by more recent musicians (*consonantias recentiorum Musicorum*) and selecting mean, apogeal, or perigeal distances to the planets according to what best fits the harmonies. The fundamental constraints seem to be that the spheres of the Sun, the fixed stars, and the Empyreum are equally spaced, and that the sphere of Saturn is halfway to the Empyreum.

In general, Bettini's distances are evidently calculated from the harmonic proportions. (No suitable harmony is available for Jupiter, given the above constraints, so it is arbitrarily placed a tone away from Mars.) The specifications of apogee, perigee, or mean distance seem to imply some kind of independently determined values for the distances. These need not have been close to the values given above, however; Bettini's designation of the distance to Jupiter as "perigeal"

---

[7] Ibid., vol. 2, p. 527. I have corrected the distance to the crystalline sphere, which is printed as 2574¾.

[8] See chapters 3–5 of van Helden, *Measuring the Universe* (Chicago, 1985).

TABLE 10.1.
Harmonic Distances According to Mario Bettini

1 Planet	2	3 Distance	4 Musical intervals
A	Earth	0	
D	Moon, at apogee	381⅙	AB:BD=9:8 (tone)
E	Venus, at perigee	687	AB:BE=5:4 (ditone) AB:AE=5:1 (double octave + ditone)
F	Mercury, at perigee	858¾	AB:BF=4:3 (fourth) AB:AF=4:1 (double octave)
C	Sun, at mean distance	1,145	AB:BC=3:2 (fifth) AB:AC=3:1 (octave + fifth)
H	Mars, at mean distance	1,374	AB:BH=5:3 (major sixth) AB:AH=5:2 (octave + ditone)
I	Jupiter, at perigee	1,603	AB:BI=3435:1832 but BI:BH=8:9 (tone)
G	Saturn, at perigee	1,717½	AB:BG=2:1 (octave) AB:AG=2:1 (octave)
K	Fixed stars	2,290	AB:BK=3:1 (octave + fifth) AB:AK=3:2 (fifth)
L	Crystalline sphere	2,576¼	AB:BL=4:1 (double octave) AB:AL=4:3 (fourth)
B	Empyreum	3,435	

could mean simply that 1,603 terrestrial radii is less than anybody else's estimate of the distance to Jupiter.

Perhaps the most interesting thing about Bettini's harmonic theory of distances is the politely dismissive attitude of his colleague toward it. Riccioli observes that the Sun is assigned the interval of a fifth and that, indeed, the fifth note in Guido's solmization of the scale is *sol*. Beyond this attempt at humor, he has little good to say about Bettini's scheme. It is, he is willing to concede, more pleasing than the Pythagorean or Platonic harmonies of the planets, but it is simply incorrect. The lunar distance is very bad, for if it were true there could be no eclipses. Jupiter's and Saturn's equations of center imply, as Copernicus showed, that they are respectively five and ten times as distant as the Sun, but Bettini has placed them much closer to the Sun. Moreover, the stars at a distance of 2,290 terrestrial radii would show a horizontal parallax of $\arcsin(1/2,290)$, about $1\frac{1}{2}$ minutes of arc, which is not observed to be true.[9]

Such criticisms reveal the Tychonic influence on Riccioli's thought. In particular, the argument from a planet's equation of center to its distance originated, as Riccioli notes, in Copernicus, and is meaningless in any geocentric system except Tycho's.

After describing and disposing of Bettini's harmonic astronomy, Riccioli turns to that of Kepler. He reproduces Kepler's table of extreme distances from book 5, chapter 4 (table 9.1, above) and paraphrases Kepler's summary of it: there is no harmony in the extreme distances of any single planet, aside from the imperfect ones for Mars and Mercury, but if one compares the distances of adjacent planets, "a harmonic law here and there shines." Kepler, however, denies that harmonies are to be found in the distances. Riccioli himself concludes that the true opinion, not differing greatly from that of Kepler, is that planetary distances cannot be obtained from harmonic proportions or musical intervals.[10]

Kepler, he continues, believes that harmony pertains necessarily to motions and not to any immobile quantity. Here Riccioli agrees, but makes a principled objection to Kepler's application of any of this to astronomy. Even if there are harmonic proportions in the intervals or motions of the planets—which he does not concede—still, harmony is itself a quality, and not something to be sought among quantities. Riccioli mocks the very notions of both *musica humana* and *musica mundana*. After all, he asks, why should harmonic proportions be intrinsic to the structure of the heavens? Quantities in the heavens are always changing, so that harmonies among them would occur rarely if ever.[11]

Even with such fundamental differences of opinion, Riccioli continues to summarize book 5 of the *Harmonice mundi*, chapter by

[9] Riccioli, *Almagestum Novum*, p. 528.
[10] Ibid.
[11] Ibid., p. 530.

chapter. He quotes Kepler's statement of the period-distance relation very accurately, reproduces several of the tables from book 5, and aptly describes Kepler's chapter 9 as "very long." He points out that chapter 10 is, among other things, an attempt to confirm the immobility of the Sun and the motion of the Earth, "without which most of the harmonies perish." Riccioli's final verdict—it has proven an enduring one— is that the celestial harmonies described by the ancients are to be understood as analogy and metaphor, and that the evidence by which Kepler and others have sought to establish those harmonies seems poetical or oratorical rather than philosophical.[12]

KEPLER THE MYSTIC?

Before closing I would like to address briefly an assertion about Kepler that has become almost commonplace: that the *Mysterium* and the *Harmonice mundi* reveal a "mystical" side to his scientific personality. This assertion arises—in all cases, it seems to me—either from sloppy thinking or from a wish to enroll Kepler among the critics of rationalism. Kepler was not a mystic. He was, undeniably, an astrologer; but in that age astrology was not yet entirely irrational. His own theories about why astrology worked (and he did believe that it worked) rested on the same theoretical foundations in geometry as his theories on the harmonies of the world. The latter, which have been the subject of this book, were meticulously rational in their development. Their origin was no more mystical than a sincere belief in God, a belief that certainly was neither abnormal nor irrational at the beginning of the seventeenth century.

The material at the end of the previous chapter might seem less than an ideal supporting environment for these statements. The very last page and a half of the *Harmonice mundi* was, indeed, presented as a dream, certainly a favorite form of mystical literature. This was a joke: Kepler did not expect his reader to believe that he had quaffed from an ancient Greek vessel, fallen asleep, and been enlightened about extraterrestrial life. He wrote in imitation of classical models—Cicero's *Dream of Scipio* and the Vision of Er from the end of Plato's *Republic*— and in a note in the margin he said so. In fact, the existence of intelligent beings on the other planets probably was something that he

---

[12] Ibid., pp. 532–33.

believed, and for the reasons he stated: the Creator certainly had the power, and probably the will, to populate the great globes he had launched in a harmonic dance around the Sun. Certainly Kepler also believed that a Mind or Minds resided in the Sun, perceiving and enjoying the intricate harmonies that had been hidden from Earth-dwellers for centuries, awaiting the perfection of astronomy.

He believed these things for ordinary reasons, down-to-earth reasons, if the reader will excuse the expression. They were not esoteric truths revealed to him; they were real possibilities, speculations about the natural world, which Kepler judged likely to be true. An early and important part of his astronomical career was the realization that the Earth moved. Along with this realization came the need to give up the illusion that the Earth occupied a privileged location, and the concomitant need to adjust his reasoning to account for his noncentral location. This adjustment is sometimes termed the "Copernican revolution in thought," and its reverberations continue to the present day, as the viewpoint of science becomes relentlessly less subjective.[13]

For Kepler, the adjustment of his viewpoint to a noncentral location quickly became routine in mathematical astronomy. In other fields, less precise and more closely tied to the unexamined notions of everyday life, it must have required greater effort. We have ample evidence of the thought Kepler devoted to these more general implications of the Copernican revolution. One encounters, scattered throughout his writings, the phrase *terram incolentes*, "dwelling on the Earth," as a description of us with our earthbound point of view. His little book of proto–science fiction called the *Somnium* ("Dream") was, among other things, an exercise in adopting other points of view in astronomy. It is thus not at all surprising that Kepler speculated about inhabitants of the Sun and planets—especially after he had discovered Sun-centered harmonies.

Dreamy speculation about life on other worlds was, let us note, no part of Kepler's scientific theories about planetary harmony. Those theories were, if anything, obsessively rational. The difficulty of his task forced Kepler to analyze the motions in greater and greater detail, until finally even he asked, as we have seen, "whether the highest creative wisdom was occupied with such trivial logic-chopping." The only reply he could find was that perhaps simpler reasons, which he

[13] This argument is, I think, at the core of C. Gillispie's stimulating essay *The Edge of Objectivity* (Princeton, 1960).

had been unable to see, lay hidden in Nature.[14] Kepler spent much of his life trying to reconstruct the *reasoning* behind the plan used by the Architect of the Heavens. He professed no knowledge of any truth, other than religion, that transcended that reasoning.

KEPLER'S ENDURING POPULARITY

The five books of the *Harmonice mundi* contain some original mathematics; a relatively small amount of astronomical theory that will always be recognized as good and important science; and a great deal of material that we, several centuries later, can only classify as valiant but misguided application of reason. Kepler's premises about the natural world are no longer viable, although that may not be the most telling basis for our unease with his harmonies. The increased scope and precision of our knowledge of that world have simply undermined the empirical supports that he first constructed, and across which he then laboriously stretched his harmonic inspiration. We judge, with the clarity of nearly four centuries' wisdom, that Kepler was wrong. No one today can reasonably believe that comparing the apparent velocities of the six naked-eye planets with the proportions deemed harmonic in Renaissance music theory was on the right track. What, then, accounts for the continuing interest in Kepler's polyhedral and harmonic theories?

Having gone this far, and perhaps feeling warmed by my over-generous draught from Kepler's cup, I will address this question briefly. It seems to me that the widespread, if diffuse, interest today in Kepler's polyhedral and harmonic theories originates in a wish that he had in some way been right. Our detailed knowledge of the world forbids us from imagining that the heavens embody harmony in the usual, everyday sense, and it is natural that sometimes we resent our inability to imagine this. Living in a physical world that has become so rational it can be described only with advanced mathematics, a world that knows no recognizable purpose, we are intrigued by the thought that a scientist like Kepler pursued a truth beyond reason. Alas, the intriguing thought is chimerical, at least in this instance. Kepler was a

[14] *G.W.* 6.342.36–41, regarding prop. 26 of book 5, chap. 9, the proposition in which the proper proportions were deduced for the apparent motions of the pair of planets entrusted with changing the genus of the celestial harmony.

man of his time: a devout student of a world acted on by little-understood forces such as magnetism, gravitation, and the astrological aspects between planets. He believed that all these forces could be understood with the help of reason.

Kepler was not a mystic, but by all accounts he was an admirable person. As a scientist, he was highly intelligent, hard-working, and relentlessly honest in subjecting his theories to empirical test. As a private person, he is reported to have been outwardly modest, good-humored, and tolerant, in another century when modesty and tolerance were far from universal. He had the bad fortune to be born at a time when religious differences were tearing Europe apart. He had the good fortune to be born at a time when it was still possible to believe, honestly, what he most earnestly wanted to believe: that the Creator of the world had employed a Design, and that he could by himself discover that Design.

Bibliography

Apelt, E. F. *Johann Keppler's Astronomische Weltansicht*. Leipzig, 1849.

Barker, A. *Greek Musical Writings II: Harmonic and Acoustic Theory*. Cambridge: Cambridge University Press, 1989.

Beer, A., and P. Beer, eds. *Kepler: Four Hundred Years*. Vol. 18 in *Vistas in Astronomy*. Oxford: Pergamon Press, 1975.

Boethius. *De institutione musica*. Ed. Giovanni Marzi. Rome: Istituto Italiano per la Storia della Musica, 1990.

Brahe, Tycho. *Astronomiae Instauratae Progymnasmata*. Prague, 1602. Facsimile reprint. Brussels: Editions Culture et Civilisation, 1969.

―――. *Opera Omnia*. Ed. J.L.E. Dreyer. Copenhagen, 1913–29. Reprint. Amsterdam: Swets and Zeitlinger, 1972.

Capella, Martianus. *De nuptiis Philologiae et Mercurii*. Ed. A. Dick. Leipzig: Teubner, 1925.

Caspar, Max. *Kepler 1571–1630*. New York: Collier, 1962.

Censorinus. *De die natali*. Ed. F. Hultsch. Leipzig, 1867.

Clavius, Christopher. *Geometria practica*. Rome, 1604.

Copernicus, Nicholas. *De revolutionibus orbium coelestium libri vi*. Nuremburg, 1543. Facsimile reprint of Kepler's own copy. New York: Johnson Reprint Corporation, 1965.

Cornford, F. M. *Plato's Cosmology: The Timaeus of Plato Translated with a Running Commentary*. New York: Harcourt, Brace, 1937.

Dickreiter, Michael. *Der Musiktheoretiker Johannes Kepler. Neue Heidelberger Studien zur Musikwissenschaft, Band 5*. Bern and Munich: Franke Verlag, 1973.

Dreyer, J.L.E. *A History of Astronomy from Thales to Kepler*. New York: Dover, 1953.

Field, J. V. *Kepler's Geometrical Cosmology*. Chicago: University of Chicago Press, 1988.

―――. "Kepler's Rejection of Solid Celestial Spheres." *Vistas in Astronomy* 23 (1979): 207–11.

―――. "Kepler's Star Polyhedra." *Vistas in Astronomy* 23 (1979): 109–41.

Fludd, Robert. *Utriusque Cosmi maioris scilicet et minoris metaphysica, physica atque technica historia*. Oppenheim, 1617–18.

Gillispie, Charles Coulston. *The Edge of Objectivity*. Princeton: Princeton University Press, 1960.

Gingerich, Owen, and R. S. Westman. "The Wittich Connection: Conflict and Priority in Late Sixteenth-Century Cosmology." *Transactions of the American Philosophical Society* 78, no. 7 (1988).

Goldstein, B. "The Arabic Version of Ptolemy's Planetary Hypotheses." *Transactions of the American Philosophical Society*, n.s., 57, no. 4 (June 1967): 3–12.

Grafton, A. "Michael Maestlin's Account of Copernican Planetary Theory." *Proceedings of the American Philosophical Society* 177, no. 6 (1973): 523–50.

Halma, ed. *Hypothèses et époques des planètes, de C. Ptolémée*. Paris, 1820. Includes the Canobic Inscription, on pp. 57–62.

Hamilton, N. T., N. M. Swerdlow, and G. J. Toomer. "The Canobic Inscription: Ptolemy's Earliest Work." In J. L. Berggren and B. R. Goldstein, *From Ancient Omens to Statistical Mechanics: Essays on the Exact Sciences Presented to Asger Aaboe*, pp. 55–73. Copenhagen: University Library, 1987.

Heath, Thomas. *Aristarchus of Samos*. Oxford: Clarendon Press, 1913. Reprint. New York: Dover, 1981.

Heiberg, J. L., ed. *Claudii Ptolemaei opera quae extant omnia*. Vol. 2. Leipzig: Teubner, 1907. Vol. 2, *Opera astronomica minora*, contains, among other works, the Canobic Inscription on pp. 149–55.

Jardine, N. *The Birth of History and Philosophy of Science*. Cambridge: Cambridge University Press, 1984.

Kepler, Johannes. *Harmonices mundi Libri v*. Linz, 1619. Facsimile edition. Brussels: Culture et Civilisation, 1968.

———. *Harmonies [sic] of the World*. Trans. C. G. Wallis with notes by Eliot Carter. In vol. 16 of *Great Books of the Western World*. Chicago: Encyclopedia Brittanica, 1952.

———. *Johannes Kepler Gesammelte Werke*. Ed. Max Caspar et al. Munich: C. H. Beck, 1939–. Cited throughout as *G.W.*, with volume, page, and, where appropriate, line numbers given. The *Harmonice mundi* is in vol. 6.

———. *Joannis Kepleri astronomi Opera Omnia*. Ed. C. Frisch. 8 vols. Frankfurt: Heyder and Zimmer, 1858–91. Cited throughout as *O.O.* The *Harmonice mundi* is in vol. 5.

———. *Kepler's Dream*. Ed. J. Lear. Trans. P. F. Kirkwood. Berkeley and Los Angeles: University of California Press, 1965.

———. *Prodromus dissertationum cosmographicarum*. Tubingen, 1596.

———. *The Secret of the Universe*. Trans. A. M. Duncan with notes by E. J. Aiton. New York: Abaris Books, 1981.

Koyré, Alexandre. *The Astronomical Revolution*. Ithaca: Cornell University Press, 1973. Trans. of *La révolution astronomique*. Paris: Hermann, 1961.

Macrobius. *Commentary on the Dream of Scipio*. Trans. with an intro. and notes by William H. Stahl. New York: Columbia University Press, 1952.

Meyer-Baer, Kathi. *Music of the Spheres and the Dance of Death*. Princeton: Princeton University Press, 1970.

Moyer, Ann E. *Musica Scientia: Musical Scholarship in the Italian Renaissance*. Ithaca: Cornell University Press, 1992.

Neugebauer, O. *A History of Ancient Mathematical Astronomy*. Berlin and New York: Springer-Verlag, 1975.

Offusius, Jofrancus. *De divina astrorum facultate in larvatam astrologiam.* Paris, 1570.

Palisca, Claude V. *Humanism in Italian Renaissance Musical Thought.* New Haven: Yale University Press, 1985.

Pauli, W. "The Influence of Archetypal Ideas on the Scientific Theories of Kepler." In *The Interpretation of Nature and the Psyche.* New York: Pantheon Books, 1955.

Plato. *The Collected Dialogs.* Ed. E. Hamilton and H. Cairns. Princeton: Princeton University Press, Bollingen Series no. 71, 1961.

Pliny. *Natural History.* Loeb edition. Revised. Cambridge: Harvard University Press, 1949.

Ptolemy, C. *Almagest.* Trans. and ed. by G. Toomer. New York: Springer-Verlag, 1984.

Randel, D., ed. *New Harvard Dictionary of Music.* Cambridge: Harvard University Press, 1986.

Riccioli, Ioanne Baptista. *Almagestum novum.* Bologna, 1651.

Russell, J. L. "Kepler's Laws of Planetary Motion: 1609–1666." *British Journal for the History of Science* 2, no. 5 (1964): 1–24.

Stephenson, B. *Kepler's Physical Astronomy.* New York: Springer-Verlag, 1987.

Swerdlow, N. M. "Pseudodoxia Copernicana: *Or, Enquiries into Very Many Received Tenents and Commonly Presumed Truths, Mostly Concerning Spheres.*" *Archives internationales d'histoire des sciences* 26 (1976): 141–55.

———, and O. Neugebauer. *Mathematical Astronomy in Copernicus's De Revolutionibus.* Berlin and New York: Springer-Verlag, 1984.

Taub, L. C. *Ptolemy's Universe.* Chicago: Open Court, 1993.

Taylor, A. E. *A Commentary on Plato's Timaeus.* Oxford: Clarendon Press, 1928.

Thoren, V. *The Lord of Uraniborg.* Cambridge: Cambridge University Press, 1990.

Van Helden, A. *Measuring the Universe.* Chicago: University of Chicago Press, 1985.

Von Jan, C. "Die Harmonie der Sphären." *Philologus* 52 (1893): 13–37.

Walker, D. P. "Kepler's Celestial Music." In *Studies in Musical Science in the Late Renaissance*, pp. 34–62. London: Warburg Institute, 1978. Reprinted from *Journal of the Warburg and Cortauld Institutes* 30 (1967): 228ff.

Weiss, Roberto. *Medieval and Humanist Greek.* Padua: Editrice Antenore, 1977.

Wilson, C. *Astronomy from Kepler to Newton.* London: Variorum, 1989.

———. "Horrocks, Harmonies, and the Exactitude of Kepler's Third Law." *Studia Copernicana* 16 (1978). Reprinted in *Astronomy from Kepler to Newton.* London: Variorum, 1989.

Index

adulterated consonances, 147n, 211n, 214

Aiton, E. J., 88n

Apelt, E. F., 167n

apparent planetary motions, 148–237

approximations assuming small eccentricity, 139–44, 227–28

Aristotle, 16, 111, 237

Aristoxenus, 22–23

astrology, 6, 76, 116, 241, 249; of Offusius, 47–63; of Ptolemy, 35–37, 116–17

Astronomia nova, 99, 137, 141, 243

Baer, N., 98

Balmer series, 11

Barlaam of Seminara, 100, 103n

al-Battānī, 69–70

Bettini, M., 246–48

Bianchi, V., 240–41

al-Bitrūjī, 53

Bode's law, 10

Boethius, 41–45

Brahe, T., 47, 73–74, 86, 96–99, 153, 226, 248

Bruce, E., 90–93

Canobic Inscription, 13, 29–32, 106

Capella, M., 25

Carter, E., 167n

Caspar, M., 11–12, 88n, 90n, 93n, 129, 167, 185–86, 220n

Censorinus, 24–25, 246

chords, Kepler's planetary, 90–97, 170–84

Cicero, 38–40, 128, 249

Clavius, C., 225

clef signatures, 90n, 166–67

comma, defined, 119–20

concinna intervals, defined, 119–20

consonant intervals, 119–20, 145–54,

170–84, 192–98; adulterated, 147n, 211n, 214; defined, 119–20

converging harmonies or motions, defined, 140

Copernicus, N.: cited by Offusius, 48–49, 51, 53; cited by Riccioli, 248; planetary distance models of, 70–73, 84–85, 113

Creator, 125–27

cube. *See* polyhedra, regular

delays, above the horizon, 56–57; in equal arcs, 136–38, 145–46

Dickreiter, M., 13, 120, 166n, 167n, 168n

diesis, defined, 119–20

diezeugmenon, 27, 104

distances to planets, 64–74; calculated from harmonies, 222–26; empirical basis of in Copernican cosmology, 70–73, 135; extreme, 146–48; for Macrobius, 40–41; for Offusius, 51–56; for Ptolemy according to Kepler, 108–10

diurnal arcs, 145–46

diverging harmonies or motions, defined, 140

dodecahedron. *See* polyhedra, regular

Dreyer, J.L.E., 12, 167n

durus. *See* genus

eccentricities of planetary orbits: according to Offusius, 53; approximations for small, 139–44, 227–28; archetypal origin of, 185–236; calculation of from harmonic principles, 220–30; in polyhedral hypothesis, 83–88

eiusdem plagae, motions of, 152–53, 201–4

encyclopedists, 38–43

equinus, 134–35, 219–20

257

Milton Keynes UK
Ingram Content Group UK Ltd.
UKHW020821250224
438379UK00009B/1021